T0255551

Mathématiques et Applications

Directeurs de la collection :
M. Hoffmann et V. Perrier

84

More information about this series at http://www.springer.com/series/2966

MATHÉMATIQUES & APPLICATIONS

Comité de Lecture 2018–2021/Editorial Board 2018–2021

Ciprian G. Gal • Mahamadi Warma

Fractional-in-Time Semilinear Parabolic Equations and Applications

Ciprian G. Gal
Department of Mathematics
Florida International University
Miami, FL, USA

Mahamadi Warma
Department of Mathematical Sciences
George Mason University
Fairfax, VA, USA

ISSN 1154-483X ISSN 2198-3275 (electronic)
Mathématiques et Applications
ISBN 978-3-030-45042-7 ISBN 978-3-030-45043-4 (eBook)
https://doi.org/10.1007/978-3-030-45043-4

Mathematics Subject Classification: 35R11, 35A01, 35A02, 35B45, 35B65, 35K58, 35E20, 35M13, 35Q86, 35Q91, 35Q92

This Springer imprint is published by the registered company Springer Nature Switzerland AG.
The registered company address is: Gewerbestrasse 11, 6330 Cham, Switzerland

Preface

This research monograph is motivated by problems in mathematical physics that involve fractional kinetic equations. These equations describe transport dynamics in complex systems which are governed by anomalous diffusion and non-exponential relaxation patterns. Such fractional equations are usually derived asymptotically from basic random walk models; among them we quote the fractional Brownian motion, the continuous-time random walk, the Lévy flight, the Schneider-Grey Brownian motion, and, more generally, random walk models based on evolution equations of single and distributed fractional order in time and/or in space (see Appendix C, for a description of the literature). Although there exists some mathematical theory about the solvability of nonlinear parabolic fractional kinetic equations, of the kind introduced in the monograph, none of the known results could be used to prove global existence and global regularity of solutions for these systems in a unifying and systematic fashion. In this situation, we gave proofs under quite transparent conditions on the nonlinearities involved, for a large class of unbounded operators that typically arise in the applications (see Chap. 1, for a complete overview). Afterwards, it turns out that these methods can be generalized and applied to reaction–diffusion systems of fractional kinetic equations. Among them, we have considered the fractional Volterra–Lotka model and the fractional nuclear model, as canonical examples that arise in population dynamics and nuclear dynamics. For the time being, the subject is clearly still open with many issues that remain still unresolved (see Sect. 1.3 and Chap. 5, for further discussions).

The reader is assumed to be familiar with the elementary theory of differential equations (such as the application of fixed-point theorems to parabolic problems) and have some basic understanding of the Lebesgue integration theory. A working knowledge of nonlinear PDE theory may be helpful, but it is not absolutely required. In this regard, the monograph may be used as a first graduate course on fractional differential equations for students in the mathematical and physical sciences. The stimulation of additional work by researchers in all scientific disciplines is also one of our other objectives.

Miami, FL, USA Ciprian G. Gal
Fairfax, VA, USA Mahamadi Warma

Acknowledgements

Until the finalization of the typescript in the present form, the authors have gratefully received invaluable comments from colleagues at many different academic institutions and research centers around the world. We give a special mention to Dr. Francisco Bersetche (University of Buenos Aires, Argentina) and Dr. Gabriel Acosta (University of Buenos Aires, Argentina). We would want to thank Professor Francesco Mainardi (Department of Physics, University of Bologna, Italy) who was the first to recommend the paper in ResearchGate.

The first-named author would like to express his appreciation for the kind hospitality and support by the University of Puerto Rico, Rio Piedras Campus, during his many short-term visits to the Department of Mathematics throughout the tenure of this book project. The first-named author would also like to thank his wife, Dr. Nadia Gal (University of Miami, Miami, USA), for her constant support and encouragement during the making of the project.

The second-named author would like to thank his wife Koumba Sankara for her support and especially to his daughter Sherry Warma for her understanding that Daddy needed to work hard at this time and therefore did not have a lot of time to play. The second-named author would also want to thank Dr. Carlos Lizama (Universidad de Santiago de Chile) and Dr. Valentin Keyantuo (University of Puerto Rico, Rio Piedras Campus) for introducing the topic of fractional calculus to him a few years ago. Finally, the second-named author would like to thank the US Air Force Office of Scientific Research (AFOSR) for funding several of his research projects on control and dynamics of fractional partial differential equations and its applications.

Acknowledgements

About This Book

We consider fractional kinetic equations characterized by the presence of a nonlinear time-dependent source, generally of arbitrary growth in the unknown function, a time derivative in the sense of Caputo, and the presence of a large class of diffusion operators. Besides classical examples involving the Laplace operator, subject to standard (namely Dirichlet, Neumann, Robin, dynamic/Wentzell, and Steklov) boundary conditions, our framework includes also nonstandard diffusion operators of "fractional" type subject to appropriate boundary conditions. We aim to give a unified scheme and analysis for the existence and uniqueness of strong and mild solutions and then deal separately with the global regularity problem. Then we extend the analysis to systems of fractional kinetic equations that include prey–predator models of Volterra–Lotka type and chemical reactions models, all of them containing possibly some fractional kinetics.

Contents

Chapter 1
Introduction

In this monograph we address some topics related to the well-posedness problem (in the sense of Hadamard) of nonlocal partial differential equations, which, in the context of mathematical physics, are also often referred to as fractional in time parabolic equations. These are known to possess solutions that exhibit anomalous behaviors [27, 28].

The nonlocal in time problem may be formulated roughly as follows. Consider an evolving nonlinear system, either described in terms of a partial or ordinary nonlocal differential equation, containing possibly some fractional kinetics. In brief, fractional kinetic equations occur often in the description of transport dynamics in complex systems which are governed by anomalous diffusion and/or non-exponential relaxation patterns. These equations are generally derived asymptotically from basic random walk models by various approaches in probability theory (see Appendix C). Given a time interval $0 < t \leq T$, and an initial state u_0, the goal is to determine whether the corresponding initial-boundary value problem (or initial-exterior value problem) can be solved globally and uniquely for any time $T > 0$. More precisely, the problem reads[1] for $0 < \alpha \leq 1$,

$$\partial_t^\alpha u = Au + f(x, t, u) \text{ in } \mathcal{X} \times (0, T], \ u(\cdot, 0) = u_0 \text{ in } \mathcal{X}. \tag{1.0.1}$$

Here $\mathcal{X}$ stands for an appropriate physical domain and f is a nonlinear, possibly time-dependent source, that is generally of arbitrary growth in the unknown function u. The precise assumptions regarding the operator A and the space $\mathcal{X}$ shall be stated precisely in Chap. 2. More concrete examples of operators and spaces $\mathcal{X}$ that enter in our framework will be given in Sect. 2.3. Finally, we shall introduce the classes of admissible nonlinearities f in Sect. 3.1. A preliminary discussion regarding the nonlinearity f will be done in Sect. 1.2.

[1] The fractional derivative ∂_t^α is meant in a generalized Caputo sense, see Sect. 2.1.

C. G. Gal, M. Warma, *Fractional-in-Time Semilinear Parabolic Equations and Applications*, Mathématiques et Applications 84,
https://doi.org/10.1007/978-3-030-45043-4_1

1.1 Historical Remarks

This is a classical problem in the theory of differential equations for $\alpha = 1$ and the scientific literature on its global solvability is quite large when $\alpha = 1$. We refer for instance to the book of Henry [19], Arendt et al. [5], Engel and Nagel [17] and Cholewa and Dlotko [10] for an introduction to this topic. There has been intensive research in this area for the ordinary problem when $\alpha \in (0, 1)$ and $A = 0$ in the last two decades (see Agarwal et al. [1], Kilbas et al. [21], Kiryakova [22]). The survey paper [1] collects a sufficiently large number of results established up to 2008, that give sufficient conditions for the (local) solvability of the ordinary problem subject also to local, nonlocal and integral initial conditions. Multivalued versions of these problems, that include fractional-in-time ordinary differential inclusions, are also considered. In the specific context of partial differential equations for (1.0.1), when A is an unbounded operator and $\alpha \in (0, 1)$, there has been only little success to address the global solvability problem in a satisfactory manner in view of the many applications that this important area holds (see Appendix C).

In this monograph, it is our main goal to focus on this latter, more difficult cases and aim to place this theory on solid footing by devising a unified approach and by giving a complete solution to the above global solvability problem as well as the global regularity problem. However, one still has to recall some pertinent literature and describe any attempts at giving a successful solution to the solvability problem, in various special cases for the nonlocal problem (1.0.1). The nonautonomous problem ($f = f(t)$) with a second-order differential (possibly quasilinear) operator A in divergence form and a Riemann-Liouville fractional derivative D_t^α has been considered by Bazhlekova [6]. Among the most important results of [6] are maximal regularity results based on two distinct approaches; one that is based on the L^p (L^q)-regularity for the corresponding linear problem and, another that exploits the theory of sums of accretive operators in a Hilbert space setting. We should point out that one distinctive disadvantage of considering a parabolic equation with a Riemann-Liouville fractional derivative is that it needs to consider unpleasant initial conditions that are generally nonlocal in nature (see also [7]). The nonautonomous problem (1.0.1) ($f = f(t)$) when A is roughly the Laplacian, but with a Caputo-like derivative ∂_t^α, is also considered by Eidelman and Kochubei [16]. They construct a fundamental solution and then investigate its various asymptotic properties. The latter become important, for instance, to establish the optimal (namely, polynomial) decay properties of solutions to problem (1.0.1), when $A = \Delta$ and $f = 0$ (see Kemppainen et al. [20] and Vergara and Zacher [35]). Mostly for the same operator but with a semilinear $f = f(u)$, the problem of blow-up in finite time of some solutions and some criteria of stability-instability are developed further in Vergara and Zacher [36], whereas some a priori bounds for some related singular evolutionary partial integro-differential equations are given in Vergara and Zacher [34]. Besides, a more general approach to the nonautonomous problem ($f = f(t)$), that is based instead on sesquilinear forms $a(t, \cdot, \cdot)$, to which a certain diffusion operator A can be associated with, is also used by Zacher [40] to prove existence and uniqueness of weak solutions in an appropriate regularity solution class. In addition, Zacher [41] provides interior Hölder regularity estimates for (1.0.1) in the case when $A = \Delta$ and $f = f(t)$, in the spirit of the classical De Giorgi-Nash regularity theorem.

Some further progress for the abstract problem (1.0.1) has also been made by Clement et al. [11], assuming $f = f(u)$ and $\alpha \in (0, 2)$, and by analyzing (1.0.1) in continuous spaces $BUC_{1-\mu}([0, T]; Y)$, $\mu \in (0, 1)$. This approach allows solutions to have a prescribed singularity at the origin in the sense that $t^{1-\mu}\|u(t)\|_Y \to 0$, as $t \to 0^+$. The ***local existence*** of smooth solutions is established via maximal regularity results for the linear equation associated with (1.0.1), by assuming that $\alpha + \mu > 1$, for $\mu \in (0, 1)$ and $\alpha \in (0, 2)$. However, these results rely too heavily on interpolation results and abstract conditions on (A, f) (see [11, Theorem 13 and (46)-(47)]) that renders their application to specific situations difficult, if not impossible. This is valid especially in those instances when the diffusion operator A turns out to be of "fractional" type. Indeed, due to the well recognized role of fractional operators in the presence of anomalous transport behaviors in some physical phenomena (see Appendix C), it is clearly important to investigate the global solvability and regularity problems for (1.0.1) in those cases. This is for instance, the case of fractional Laplace operators $(-\Delta)^s_{\mathcal{X}}$, $s \in (0, 1)$ (see Sect. 2.3), which as we shall see, turn out to have quite different properties than the classical Laplacian. In particular, such operators are known to generally lack, with the exception of some special cases, an explicit characterization in terms of (Sobolev) function spaces for any fractional powers $Y_\theta := D((-A)^\theta)$, $\theta \in (0, 1)$. For this reason, the application of the results of [11] seems then best suited in those situations when A is a "local" operator, say a second-order quasilinear operator in divergence form (see the example of [11, Section 9]). Unfortunately, this is also the point taken by Andrade, Carvalho et al. [13] and Guswanto and Suzuki [18], who establish a local theory of mild solutions for problem (1.0.1), when $f = f(u)$ and A is a sectorial (nonpositive) operator, using the concept of the so-called θ-regular maps (see also Neto [14, Chapter 3]). The latter means that the nonlinearity f is locally Lipschitz as a mapping from $Y_{1+\theta}$ to $Y_{\gamma(\theta)}$, for some $\gamma(\theta) \in (0, 1)$, a choice which turns out to be useful in the treatment of the problem (1.0.1) with nonlinearities of critical and subcritical polynomial growth. Here criticality is meant in the sense that there exist some critical exponent q, given by some well-known continuous embedding results in Sobolev theory, that controls the polynomial growth of the function $f(u)$ as $|u| \to \infty$. However, these techniques also suffer from several drawbacks: first, they are not well suited to deal with nonlinearities that are also x-dependent ($f = f(x, u)$), and secondly, such Lipschitz conditions lack any transparency and simplicity since once again they strongly rely on the explicit characterization of Y_θ in terms of known Sobolev spaces. Indeed, most of the applications of these techniques seem to be found only in the case of classical operators, such as, when $A = \Delta$ (cf. [13, 18]). Notably, other works by Liu and Liu [25, 26], Ouahab [29], Zhang and Liu [42], Wang and Zhou [37] and Wang et al. [38] (and the references there in), have obtained similar comparable results on the local existence of mild solutions, but with conditions on the nonlinearity which are also too strong; namely, by assuming either that $f = f(u)$ is a globally Lipschitz function, or locally Lipschitz with a sufficiently small Lipschitz constant and/or a sufficiently small growth as a function of u. In the case when A is related to a fractional operator, in particular, $A = (-\Delta)^s_{\mathcal{X}}$, $s \in (0, 1)$, for a compact Riemannian manifold $\mathcal{X}$ (without boundary), some local existence results for (1.0.1) with $f = f(u)$ assuming some polynomial growth, are also contained in the lecture notes by Taylor [33]. Hölder continuity for the problem (1.0.1) assuming $f = f(t)$ and an operator A that is related to the fractional Laplacian, has been established by Allen, Caffarelli et al. [3], when ∂_t^α is meant as a (one-sided) nonlocal derivative in the sense of Marchaud.[2] Some related nonlocal ordinary differential equations associated with a Marchaud type of fractional derivative are also investigated in [8] and [24], with the latter also providing an extensive comparison between the Marchaud, Riemann-Liouville and Caputo fractional derivatives, respectively. Further applications of the framework from [24] to nonlinear time fractional PDEs are also given in [23]. Most recently, the case $1 < \alpha < 2$ is further investigated in [4], where the existence of weak and strong (energy-like) solutions in various settings are among the central results. The work in [4] offers a new and fresh alternative from the contribution in [11] in the sense that the assumptions imposed on the operator A, as well as the nonlinearity f, are once again more natural and transparent than the conditions imposed by [11]. It is worth emphasizing that fractional kinetic equations typically exhibit a variety of behaviors which are completely different from the classical case, while in fact solutions of fractional equations can sometimes be "arbitrarily" complicated, as shown in [9, 15].

1.2 On Overview of Main Results and Applications

But none of these theories address the global regularity problem for the full semilinear problem (1.0.1) in a meaningful way for practical applications. Furthermore, these theories are far from being applicable to reaction–diffusion systems with vectorial quantities $u \in \mathbb{R}^m$, $m > 1$, which contain some fractional kinetics, but which draw their breath from important applications in biology, chemistry and finance. This is particularly relevant in the context of biological systems where the mechanism is necessarily more involved and complex due to a richer structure associated to the corresponding couplings under consideration (especially when different nonlocal derivatives $\partial_t^{\alpha_i}$, with $\alpha_i \in (0, 1]$, $i \in \{1, \ldots, m\}$ are involved). As a matter of fact, understanding the connections between the right fractional parameter in concrete biological settings, also in relation to the environment, seem to be an important topic in optimization, see e.g. [32]. Interesting applications also arise in neuroscience and in neural networks, see e.g. [31]. As one knows, there is

[2] See Sect. 2.1, for further details.

an extensive literature on the topic of reaction–diffusion systems when $\partial_t^{\alpha_i} \equiv \partial_t$, for all $i \in \{1, \ldots, m\}$ (see, for instance, Yagi [39]). Although, the techniques and methodologies developed in this monograph for the scalar equation (1.0.1) shall prove quite useful, the problem of global solvability and regularity for m-systems needs to be addressed directly and independently. The main tools are borrowed essentially from the methodology that deals with the scalar equation. But let us first mention our unified approach and the type of results one can obtain for the scalar equation (1.0.1), in a successful manner that also covers the existing theory for (1.0.1) when $\alpha = 1$ and/or A is a uniformly elliptic (second-order) operator. Although we cannot give a complete review of the literature for the problem (1.0.1) when $\alpha = 1$ and $A = \Delta$, the lecture notes of Rothe [30] give a good account of the main developments concerning the global existence of solutions for semilinear parabolic equations in that case.

The present work is concerned with some fundamental questions for the initial-boundary (or initial-exterior) value problem (1.0.1), namely,

- the global existence of non-regular (mild) solutions;
- the existence of sufficiently smooth (strong) solutions, for which the nonlocal equation is satisfied pointwise in time;
- the global regularity problem, to establish sufficient conditions and uniform a priori bounds in such a way that each non-regular solution becomes a global smooth solution on $(0, \infty)$, and
- what happens to any global solution of (1.0.1) as $\alpha \to 1$?

These aspects are studied in a unified framework for the scalar equation (1.0.1) in a first part, and then for general nonlocal reaction–diffusion m-systems, and some of their applications, in the second part. Of course, in this part only some non-trivial examples, which are motivated by applications in mathematical biology and chemistry, will be investigated thoroughly. But the general setting developed here allows to derive similar results for other important nonlocal reaction–diffusion m-systems which can be handled by the same techniques.

In our unified framework, the essential assumption about the semigroup $S(t)$, associated with the diffusion operator A, is an ultracontractivity estimate of the form

$$\|S(t)\|_{\mathcal{L}(L^p(\mathcal{X}),L^q(\mathcal{X}))} \leq Ct^{-\beta_A\left(\frac{1}{p}-\frac{1}{q}\right)}, \ \forall t > 0, \tag{1.2.1}$$

for $1 \leq p \leq q \leq \infty$, and some positive constant $C = C(p, q, \mathcal{X}) > 0$. The constant $\beta_A > 0$ is assumed independent of p, q; it plays the role of capturing the degree of "smoothness" of the fundamental solution associated with the diffusion operator A. For symmetric semigroups, it is well-known that (1.2.1) is also connected to optimal Sobolev inequalities (see, for instance, Davies [12]). For instance, when $A = \Delta$ is subject to classical boundary conditions and $\mathcal{X} \subset \mathbb{R}^N$ is a smooth bounded domain, it holds $\beta_A = N/2$, while for $A = (-\Delta)^s_{\mathcal{X}}$, $s \in (0, 1)$, we have $\beta_A = N/2s$. Condition (1.2.1) constitutes the main assumption upon which our general theory is built on. Indeed, it allows to consider a general family of diffusion operators, including the classical ones as well as ones of "fractional" type, in addition to other non-standard examples of diffusion. We refer the reader to Sect. 2.3 for many examples of operators that are covered by our framework. Next, our main assumption on the nonlinearity $f = f(x, t, u)$ is quite transparent and easy to verify in applications; in addition to other basic conditions (which imply that f is locally Lipschitz as a function of u, with the same c, Q; see (**F4**)), it is simply measurable and satisfies (as $|u| \to \infty$) a growth condition of the form

$$|f(x, t, u)| \leq c(x, t)\, Q(u), \text{ for all } u \in \mathbb{R}, \text{ a.e } (x, t) \in \mathcal{X} \times (0, \infty), \qquad (1.2.2)$$

for some (nonnegative) integrable function[3] $c \in L_{q_1,q_2}$, $1 \leq q_1, q_2 \leq \infty$. Of course, the real-valued (positive) function Q generally captures the growth of the nonlinearity as $|u| \to \infty$. In some cases, we will allow it to behave polynomially in the sense that

$$Q(u) \sim |u|^{\gamma}, \; \gamma \geq 1, \text{ as } |u| \to \infty. \qquad (1.2.3)$$

In this monograph, we look for complete results regarding the solvability of problem (1.0.1) in such a way that also the case $\alpha = 1$ is automatically included. To this end, let us define a number $W = W(\alpha, f, p_0, q_1, q_2, Q) \in \mathbb{R}$, by

$$W := \frac{\mathfrak{n}}{q_1} + \frac{1}{q_2} + (\gamma - 1)\frac{\mathfrak{n}}{p_0}, \; \mathfrak{n} := \beta_A \alpha, \; \alpha \in (0, 1],$$

as the essential range for problem (1.0.1) for which (at least local) well-posedness can be established. Theorem 3.1.4 establishes the existence of (locally-defined) mild solutions for (1.0.1) for non-regular initial data $u_0 \in L^{p_0}(\mathcal{X})$, $1 \leq p_0 \leq \infty$, in the following cases:

(i) $W \leq \alpha$, under the assumptions (1.2.2)–(1.2.3) for some $\gamma \in [1, \infty)$ and $p_0 \in [1, \infty)$.
(ii) When $\frac{\mathfrak{n}}{q_1} + \frac{1}{q_2} < \alpha$, $p_0 = \infty$ and Q is an arbitrary positive function.

The critical case, defined by the equality $W = \alpha$ in case (i), is included; this range turns out to be also optimal in the sense that for some $p_0 \in [1, \infty]$ and $\gamma \geq 1$ that satisfy $W > \alpha$, there are ***no locally-defined mild solutions*** for certain initial data $u_0 \in L^{p_0}(\mathcal{X})$ (see Chap. 5, Remark 5.0.2). In the above cases (i)–(ii), Theorem 3.1.10 proves the existence of mild solutions on a maximal interval of existence $(0, T_{\max})$, such that (non-regular) mild solutions are always locally bounded, namely, $u \in L^{\infty}((0, T_{\max}); L^{\infty}(\mathcal{X}))$. As usual the time $T_{\max} > 0$ is such that, either $T_{\max} = \infty$ or $T_{\max} < \infty$ with $\|u(t)\|_{L^{\infty}(\mathcal{X})} \to \infty$ as $t \to T_{\max}^{-}$. In other words, knowledge of the a priori bound $u \in L^{\infty}((0, T); L^{\infty}(\mathcal{X}))$, for any (fixed) time $T > 0$, is essential for both the global solvability problem and the

[3] See Sect. 3.1, for the precise definition of L_{q_1,q_2}.

global regularity problem, as we shall see in what follows. It is worth stressing out that we recover the essential range for local solvability in the case when $\beta_A = N/2$ and $\alpha = 1$ (see Rothe [30]).

A major development in the monograph is a unified theory of strong solutions that contains the case $\alpha = 1$ as a particular case. Denote by A_p the generator of the semigroup $S(t)$ on $L^p(\mathcal{X})$ so that $A_2 \equiv A$. Our next goal is to show that any maximally-defined mild solution can become, under natural conditions, a strong ***bounded*** solution on $(0, T_{\max})$; the latter is by definition a sufficiently smooth solution of the abstract equation (1.0.1) in some Banach space. In particular, Theorem 3.2.2 proves the aforementioned statement in the space $L^p(\mathcal{X})$, $1 \le p \le \infty$, for an initial datum $u_0 \in D\left(A_p\right)$, $p \in (\beta_A, \infty) \cap (1, \infty)$, under either one of the following two alternatives:

(a) if $p \ge q_1$, assume f satisfies (1.2.2) and a locally Lipschitz-Hölder condition (see **(F4)**–**(F5)**) with $q_2 \in (1/\alpha, \infty]$ and $\theta \in (\beta_A/p, 1)$ satisfying

$$\alpha(1-\theta) - \mathrm{n}\left(\frac{1}{q_1} - \frac{1}{p}\right) > \frac{1}{q_2};$$

(b) if $p \le q_1$, assume f satisfies (1.2.2) and a locally Lipschitz-Hölder condition (see **(F4)**–**(F5)**) with $q_2 \in (1/\alpha, \infty]$ and $\theta \in (\beta_A/p, 1)$ satisfying

$$\alpha(1-\theta) > \frac{1}{q_2}.$$

The conditions **(F4)**–**(F5)** are generally satisfied in practical applications, as it can be observed in the context of specific examples, and there are situations when **(F5)** can be even entirely dropped, especially when $f = f(x, u)$. Besides, we show in Theorem 3.2.6 that every bounded (maximally-defined) mild solution constructed in case (ii), becomes indeed a strong solution on $(0, T_{\max})$. The techniques exploited in Sect. 3.2 provide several important developments among which we can mention the fact that the same a priori bound $u \in L^\infty((0, T); L^\infty(\mathcal{X}))$, for the maximal strong solution, suffices for its ***global*** regularity. We refer the reader to Sect. 3.2 for further details and more precise statements of the above regularity results. Finally, in Sect. 3.3 we obtain some results on the differentiability properties of strong solutions in the case $\alpha \in (0, 1)$ (the case $\alpha = 1$ is well known, see e.g., [17, 19]). Such results are necessary in order to estimate the error in numerical approximations of the solution of (1.0.1). They turn out to be also important in existence proofs of certain energy inequalities that are used to derive the long term behavior for such solutions as time goes to infinity (see, for instance, the discussion in Sect. 3.4).

Perhaps then the next important point is the construction of uniform a priori bounds that imply the aforementioned bound in $u \in L^\infty((0, T); L^\infty(\mathcal{X}))$, which is necessary to completely solve the global regularity problem. This problem may be formulated roughly as follows. Restricting to a smooth initial datum $u_0 \in D\left(A_p\right)$, consider the corresponding (unique) strong solution of (1.0.1) whose existence is assured by one of the previous statements. The main idea is to take a weak bound of the form $u \in L^{r_1}((0,T); L^{r_2}(\mathcal{X}))$, that is known to be satisfied a priori for some $1 \le r_1, r_2 \le \infty$, and to convert this information into an explicit bound for the strong solution in $L^{\infty}((0,T); L^{\infty}(\mathcal{X}))$, for any $T > 0$. This goal will be achieved by two essentially different methods. On one hand, we shall employ and extend a "feedback" argument used by Rothe [30] in the case $\beta_A = N/2$ and $\alpha = 1$, to provide such a statement in Theorem 3.4.1. This method has the advantage that it employs only elementary inequalities and bootstrapping arguments involving only space and time integrals. The second method we use to derive such a priori estimates is based on an iterative Moser procedure that was exploited by Alikakos [2] once again in the case $\beta_A = N/2$ and $\alpha = 1$. We extend this procedure in our general setting when $\alpha \in (0, 1]$ and A is a "properly-behaved" diffusion operator that covers many of the examples we have in mind (see Sect. 2.3); one advantage of this scheme is that the estimates remain ***uniform*** as the order α of the fractional in time derivative ∂_t^{α} approaches 1. Although, the precise statements of these global estimates are somewhat more complicated to state here, we refer the reader to Sect. 3.4 for the corresponding results. Furthermore, taking into account the above developments, we do mention that we can finally address the important issue of convergence as $\alpha \to 1$ for problem (1.0.1) under quite natural conditions on (A, f). The statement of Theorem 3.5.1 (in Sect. 3.5) shows in particular that for nonlinearities that satisfy $f(x,t,u)\, u \le c_0(x,t)\left(1+u^2\right)$, for some $c_0 \in L_{\infty,\infty}$, any globally bounded mild solution $u = u_{\alpha}$ of the abstract problem (1.0.1) converges in the sense that

$$\lim_{\alpha \to 1} \sup_{t \in (0,T)} \|u_{\alpha}(t) - u_1(t)\|_{L^{\infty}(\mathcal{X})} = 0, \text{ for any } T > 0, \tag{1.2.4}$$

to a bounded mild solution u_1 of problem (1.0.1) in the case when $\alpha = 1$. Finally, one further important application of these results is that they guarantee the global solvability and regularity of solutions to a fractional in time Fischer-KPP like equation (see Sect. 3.7), for a large class of interesting diffusion operators A, as well as the aforementioned convergence result (1.2.4) holds.

1.3 Results on Nonlocal Reaction–Diffusion Systems

The setting in the first part of the monograph, which deals solely with the scalar equation (1.0.1), can be extended and applied to reaction–diffusion systems for an unknown vectorial quantity $u = (u_1, \ldots, u_m) \in \mathbb{R}^m$, $m \in \mathbb{N}$. Let $d_i = 0$ for $i = 1, \ldots, r$ and $d_i > 0$ for $i = r+1, \ldots, m$. We also allow the case $r = 0$ to occur so that all $d_i > 0$ for $i = 1, \ldots, m$. Next, let $D = diag(d_1, \ldots, d_m)$ be the diagonal matrix of diffusion coefficients and assume that $u_0 = (u_{01}, \ldots, u_{0m})(x) \in \mathbb{R}^m$, for $x \in \mathcal{X}$, models the initial data. Let $f = (f_1, \ldots, f_m)(x, t, u_1, \ldots, u_m)$ be a nonlinear function that models possible interactions between the various quantities u_i $(i = 1, \ldots, m)$. After that we can set the diagonal (matrix) operator

$A = diag(A_1, \ldots, A_m)$, where each A_i is a proper diffusion operator, and then introduce the following notation $\partial_t^\alpha u = \left(\partial_t^{\alpha_1} u_1, \ldots, \partial_t^{\alpha_m} u_m\right) \in \mathbb{R}^m$, where each nonlocal derivative $\partial_t^{\alpha_i} u_i \in \mathbb{R}$ is understood in the generalized sense of Caputo (see Sect. 2.1) for $\alpha_i \in (0,1)$, whereas for $\alpha_i = 1$, $\partial_t^1 = \partial_t = d/dt$.

Consider then the following nonlocal reaction–diffusion system

$$\partial_t^\alpha u = DAu + f(x,t,u), \ (x,t) \in \mathcal{X} \times (0,\infty), \ u_{|t=0} = u_0 \text{ in } \mathcal{X}. \qquad (1.3.1)$$

Of course, the above framework allows a general study of (1.3.1) when diffusion can be also completely ignored in some of the components of u. In this case, we are looking for sufficiently general conditions on (A, f) such that it possesses maximally-defined bounded (mild) solutions as well as strong solutions. For $u_0 \in L^\infty(\mathcal{X}, \mathbb{R}^m)$, Theorem 4.1.2 proves the existence of (maximal) bounded mild solutions under natural assumptions on f and analogous conditions on the operators A_i, as in the scalar case (1.0.1). The conditions on the nonlinearity roughly imply that f is locally Lipschitz-Hölder in bounded subsets B of $\mathcal{X} \times [0,\infty) \times \mathbb{R}^m$ (see **(SF1)**–**(SF2)**). For more regular initial data u_0, Theorem 4.1.3 even shows the existence of (maximal) unique strong solutions; these are sufficiently smooth solutions satisfying (1.3.1) pointwise in time on the maximal interval of existence. In Chap. 4, we shall introduce the classes of diagonal matrix operators A and matrix diffusion coefficients D, and give the class of admissible nonlinearities that enter in the framework of the system (1.3.1).

As it was pointed out at the beginning of the introduction, the issues of global solvability and regularity for (1.3.1) must be instead addressed independently for problems that are suggested by a practical application. To this end, our main focus now turns onto some nonlocal systems that arise in population dynamics (see Sect. 4.2) and nuclear dynamics (see Sect. 4.3). The first one is a nonlocal model of Volterra–Lotka type and consists of a coupled system for two nonlocal partial differential equations, with the "nonlocality" being expressed in both space and time, in one of the equations. The second systems consists of a coupled system similar to the first one, but one of the components satisfies instead a nonlocal ordinary differential equation (namely, diffusion is totally ignored for that component). The results developed for the scalar equation (1.0.1) in the first part of the monograph, are crucial to the investigation of global solvability and global regularity for these specialized systems. They allow to prove sharper results by applying the corresponding theorems in the first part, to the diffusion equations for each individual component u_i of u, and by treating the other remaining components u_j, $j \neq i$, in the nonlinearity f_i, as part of a special "weight" function $c(x,t)$. The advantage of this approach is that only very little information, such as some a priori L_{q_1,q_2}-bound is required for $c(x,t)$, to deduce global information on that component u_i. Then these arguments can be repeated for each component of u, one by one, until the entire range of $i \in \{1, \ldots, m\}$ is exhausted. We refer the reader to Chap. 4 for the precise statements of these results and further details.

Finally, as we said above, the coverage of particular cases of nonlocal reaction–diffusion systems in this monograph is necessarily limited. But their successful treatment proves to be considerable progress in this area especially when such feats do not seem to have been attempted before. Nevertheless, we emphasize that other (more complicated) nonlocal systems can be included and treated within the above framework; we hope that their investigation might be of interest for other researchers in this area.

The content of this monograph is as follows. In Chap. 2 we make a complete study in the context of linear nonhomogeneous equations. Chapters 3 and 4 are devoted to the main results related to the global solvability and global regularity problems for the scalar equation (1.0.1) and reaction–diffusion system (1.3.1), respectively. Chapter 5 is devoted to some open problems and future directions of research. Three different appendices are included. Appendix A includes a number of supportive technical tools. Appendix B contains a complete discussion of the properties of the regional fractional Laplace operator in a bounded domain. Appendix C gives a full account of the physical literature on fractional kinetic equations and several applications involving evolution equations of single and distributed fractional order in time and/or in space.

References

1. R.P. Agarwal, M. Benchohra, S. Hamani, A survey on existence results for boundary value problems of nonlinear fractional differential equations and inclusions. Acta Appl. Math. **109**(3), 973–1033 (2010)
2. N.D. Alikakos, L^p bounds of solutions of reaction-diffusion equations. Commun. Partial Differ. Equ. **4**(8), 827–868 (1979)
3. M. Allen, L. Caffarelli, A. Vasseur, A parabolic problem with a fractional time derivative. Arch. Ration. Mech. Anal. **221**(2), 603–630 (2016)
4. E. Alvarez, C.G. Gal, V. Keyantuo, M. Warma, Well-posedness results for a class of semi-linear super-diffusive equations. Nonlinear Anal. **181**, 24–61 (2019)
5. W. Arendt, C.J.K. Batty, M. Hieber, F. Neubrander, *Vector-Valued Laplace Transforms and Cauchy Problems*. Monographs in Mathematics, vol. 96, 2nd edn. (Birkhäuser/Springer Basel AG, Basel, 2011)
6. E.G. Bajlekova, Fractional Evolution Equations in Banach Spaces. Technische Universiteit Eindhoven Eindhoven, The Netherlands, 2001
7. E. Bazhlekova, Strict L^p solutions for nonautonomous fractional evolution equations. Math. Balkanica (N.S.) **26**(1–2), 25–34 (2012)
8. A. Bernardis, F.J. Martin-Reyes, P.R. Stinga, J.L. Torrea, Maximum principles, extension problem and inversion for nonlocal one-sided equations. J. Differ. Equ. **260**, 6333–6362 (2016)
9. C. Bucur, Local density of Caputo-stationary functions in the space of smooth functions. ESAIM Control Optim. Calc. Var. **23**(4), 1361–1380 (2017)
10. J.W. Cholewa, T. Dlotko, *Global Attractors in Abstract Parabolic Problems*. London Mathematical Society Lecture Note Series, vol. 278 (Cambridge University Press, Cambridge, 2000)
11. Ph. Clément, S-O. Londen, G. Simonett, Quasilinear evolutionary equations and continuous interpolation spaces. J. Differ. Equ. **196**(2), 418–447 (2004)
12. E.B. Davies, *Heat Kernels and Spectral Theory*. Cambridge Tracts in Mathematics, vol. 92 (Cambridge University Press, Cambridge, 1990)

13. B. de Andrade, A.N. Carvalho, P.M. Carvalho-Neto, P. Marí n Rubio, Semilinear fractional differential equations: global solutions, critical nonlinearities and comparison results. Topol. Methods Nonlinear Anal. **45**(2), 439–467 (2015)
14. P.M. de Carvalho Neto, Fractional Differential Equations: A Novel Study of Local and Global Solutions in Banach Spaces. PhD thesis, Universidade de São Paulo, 2013
15. S. Dipierro, O. Savin, E. Valdinoci, All functions are locally s-harmonic up to a small error. J. Eur. Math. Soc. **19**(4), 957–966 (2017)
16. S.D. Eidelman, A.N. Kochubei, Cauchy problem for fractional diffusion equations. J. Differ. Equ. **199**(2), 211–255 (2004)
17. K.-J. Engel, R. Nagel, *One-Parameter Semigroups for Linear Evolution Equations*. Graduate Texts in Mathematics, vol. 194 (Springer, New York, 2000). With contributions by S. Brendle, M. Campiti, T. Hahn, G. Metafune, G. Nickel, D. Pallara, C. Perazzoli, A. Rhandi, S. Romanelli and R. Schnaubelt
18. B.H. Guswanto, T. Suzuki, Existence and uniqueness of mild solutions for fractional semilinear differential equations. Electron. J. Differ. Equ. **168**, 16 (2015)
19. D. Henry, *Geometric Theory of Semilinear Parabolic Equations*. Lecture Notes in Mathematics, vol. 840 (Springer, Berlin/New York, 1981)
20. J. Kemppainen, J. Siljander, V. Vergara, R. Zacher, Decay estimates for time-fractional and other non-local in time subdiffusion equations in $\mathbb{R}^d$. Math. Ann. **366**(3–4), 941–979 (2016)
21. A.A. Kilbas, H.M. Srivastava, J.J. Trujillo, *Theory and Applications of Fractional Differential Equations. North-Holland Mathematics Studies*, vol. 204 (Elsevier Science B.V., Amsterdam, 2006)
22. V. Kiryakova, *Generalized Fractional Calculus and Applications*. Pitman Research Notes in Mathematics Series, vol. 301 (Longman Scientific & Technical, Harlow); copublished in the United States with John Wiley & Sons, Inc., New York, 1994
23. J.-G. Li, L. Liu, Some compactness criteria for weak solutions of time fractional PDEs. SIAM J. Math. Anal. **50**, 3693–3995 (2018)
24. J.-G. Li, L. Liu, A generalized definition of caputo derivatives and its application to fractional ODEs. SIAM J. Math. Anal. **50**, 2867–2900 (2018)
25. X. Liu, Z. Liu, Existence results for fractional semilinear differential inclusions in Banach spaces. J. Appl. Math. Comput. **42**(1–2), 171–182 (2013)
26. X. Liu, Z. Liu, On the 'bang-bang' principle for a class of fractional semilinear evolution inclusions. Proc. R. Soc. Edinb. Sect. A **144**(2), 333–349 (2014)
27. F. Mainardi, Fractional calculus: some basic problems in continuum and statistical mechanics, in *Fractals and Fractional Calculus in Continuum Mechanics (Udine, 1996)*. CISM Courses and Lectures, vol. 378 (Springer, Vienna, 1997), pp. 291–348
28. B.B. Mandelbrot, J.W. Van Ness, Fractional Brownian motions, fractional noises and applications. SIAM Rev. **10**, 422–437 (1968)
29. A. Ouahab, Fractional semilinear differential inclusions. Comput. Math. Appl. **64**(10), 3235–3252 (2012)
30. F. Rothe, *Global Solutions of Reaction-Diffusion Systems*. Lecture Notes in Mathematics, vol. 1072 (Springer, Berlin, 1984)
31. F. Santamaria, S. Wils, E. De Schutter, G.J. Augustine, Anomalous diffusion in purkinje cell dendrites caused by spines. Neuron **52**(4), 635–648 (2006)
32. J. Sprekels, E. Valdinoci, A new type of identification problems: optimizing the fractional order in a nonlocal evolution equation. SIAM J. Control Optim. **55**(1), 70–93 (2017)
33. M. Taylor, Remarks on fractional diffusion equations. Preprint. http://mtaylor.web.unc.edu/files/2018/04/fdif.pdf
34. V. Vergara, R. Zacher, A priori bounds for degenerate and singular evolutionary partial integro-differential equations. Nonlinear Anal. **73**(11), 3572–3585 (2010)
35. V. Vergara, R. Zacher, Optimal decay estimates for time-fractional and other nonlocal subdiffusion equations via energy methods. SIAM J. Math. Anal. **47**(1), 210–239 (2015)
36. V. Vergara, R. Zacher, Stability, instability, and blowup for time fractional and other nonlocal in time semilinear subdiffusion equations. J. Evol. Equ. **17**(1), 599–626 (2017)

37. J.R. Wang, Y. Zhou, Existence and controllability results for fractional semilinear differential inclusions. Nonlinear Anal. Real World Appl. **12**(6), 3642–3653 (2011)
38. R.-N. Wang, D.-H. Chen, T.-J. Xiao, Abstract fractional Cauchy problems with almost sectorial operators. J. Differ. Equ. **252**(1), 202–235 (2012)
39. A. Yagi, *Abstract Parabolic Evolution Equations and Their Applications*. Springer Monographs in Mathematics (Springer, Berlin, 2010)
40. R. Zacher, Weak solutions of abstract evolutionary integro-differential equations in Hilbert spaces. Funkcial. Ekvac. **52**(1), 1–18 (2009)
41. R. Zacher, A De Giorgi–Nash type theorem for time fractional diffusion equations. Math. Ann. **356**(1), 99–146 (2013)
42. Z. Zhang, B. Liu, Existence results of nondensely defined fractional evolution differential inclusions. J. Appl. Math. **2012**, 19 (2012). Art. ID 316850

Chapter 2
The Functional Framework

We first introduce some background. Let Y, Z be two Banach spaces endowed with norms $\|\cdot\|_Y$ and $\|\cdot\|_Z$, respectively. We denote by $Y \hookrightarrow Z$ if $Y \subseteq Z$ and there exists a constant $C > 0$ such that $\|u\|_Z \leq C \|u\|_Y$, for $u \in Y \subseteq Z$. In particular, this means that the injection of Y into Z is continuous. In addition, if Y is dense in Z, then we denote by $Y \overset{d}{\hookrightarrow} Z$, and finally if the injection is also compact we shall denote it by $Y \overset{c}{\hookrightarrow} Z$. We denote by $\mathcal{L}(Y, Z)$ the space of all continuous (bounded) operators from Y to Z. If $Y = Z$, we let $\mathcal{L}(Y, Z) = \mathcal{L}(Y)$. By the dual Y^* of Y, we think of Y^* as the set of all (continuous) linear functionals on Y. When equipped with the operator norm $\|\cdot\|_{Y^*}$, Y^* is also a Banach space. Let also $\mathcal{X}$ be a (relatively) compact Hausdorff space and $\mathfrak{m}$ a Radon measure on $\mathcal{X}$ such that $\mathfrak{m}$ is supported on $\mathcal{X}$. By $\mathcal{X}$ a relatively compact Hausdorff space, we mean that there exists a metric space $\widetilde{\mathcal{X}}$ such that $\mathcal{X} \subset \widetilde{\mathcal{X}}$ and the closure $\overline{\mathcal{X}}$ (in $\widetilde{\mathcal{X}}$) of $\mathcal{X}$ is a compact set. We denote by $(\cdot, \cdot)$ the inner product in $L^2(\mathcal{X}) = L^2(\mathcal{X}, \mathfrak{m})$ and consider $L^p(\mathcal{X}) = L^p(\mathcal{X}, \mathfrak{m})$ to be the corresponding Banach space for $p \neq 2$, with norm $\|\cdot\|_{L^p(\mathcal{X})}$.

2.1 The Fractional-in-Time Linear Cauchy Problem

Before introducing the mentioned Cauchy problem, we recall the definition and some useful properties of convolutions that will be frequently used throughout the monograph. Given two measurable functions u and v defined on $(0, \infty)$, the convolution of u and v, denoted by $u * v$, is the function defined on $(0, \infty)$ and given by

$$(u * v)(t) := \int_0^t u(t - \tau)v(\tau)\, d\tau, \quad t > 0,$$

C. G. Gal, M. Warma, *Fractional-in-Time Semilinear Parabolic Equations and Applications*, Mathématiques et Applications 84,
https://doi.org/10.1007/978-3-030-45043-4_2

whenever the integral exists. If $v \in C([0,\infty))$ and $(u * v')(t)$ is well defined for $t > 0$, then for every $t > 0$,

$$\frac{d}{dt}\left[(u * v)(t)\right] = u(t)v(0) + (u * v')(t).$$

In general, let $k \in \mathbb{N}$. If $u \in C^{k-1}((0,\infty))$, $v \in C^{k-1}([0,\infty))$ and $(u * v^{(k)})(t)$ is well defined for $t > 0$, then for every $t > 0$,

$$\frac{d^k}{dt^k}\left[(u * v)(t)\right] = \sum_{j=0}^{k-1} u^{(k-1-j)}(t)v^{(j)}(0) + (u * v^{(k)})(t).$$

We refer to the monograph [11, Section 1.3] for more and precise properties of convolutions.

We next recall the notion of fractional-in-time derivative in the sense of Caputo. Let $\alpha \in (0, 1)$ and define

$$g_\alpha(t) = \begin{cases} \dfrac{t^{\alpha-1}}{\Gamma(\alpha)} & \text{if } t > 0, \\ 0 & \text{if } t \leq 0, \end{cases} \tag{2.1.1}$$

where Γ is the usual Gamma function.

Definition 2.1.1 Let Y be a Banach space, $T > 0$ and let $f \in C([0,T];Y)$, with $g_{1-\alpha} * f \in W^{1,1}((0,T);Y)$. The Riemann-Liouville fractional derivative of order $\alpha \in (0, 1)$ is given by

$$D_t^\alpha f(t) := \frac{d}{dt}\Big(g_{1-\alpha} * f\Big)(t) = \frac{d}{dt}\int_0^t g_{1-\alpha}(t-\tau) f(s)\, d\tau,$$

for almost all $t \in (0, T]$. We define the fractional derivative of order α, of Caputo-type, as follows:

$$\partial_t^\alpha f(t) := D_t^\alpha\left(f(t) - f(0)\right), \text{ a.e. } t \in (0, T]. \tag{2.1.2}$$

We notice that (2.1.2) in Definition 2.1.1 gives a weaker notion of (Caputo) fractional derivative compared to the original definition introduced by Caputo in the late 1960s (see [22]). In particular, (2.1.2) does not require f to be differentiable. In addition we have that $\partial_t^\alpha(c) = 0$, for any constant c. For this reason, (2.1.2) offers a better alternative than the classical notion of Caputo derivative. Refer to, for instance, [28, Proposition 2.34], which shows the two notions coincide when f is smooth enough, namely,

$$\partial_t^\alpha f = g_{1-\alpha} * \partial_t f, \quad \text{for } f \in C^1([0,T];Y). \tag{2.1.3}$$

For such smooth f, one may use also integration by parts in the Caputo definition (2.1.3) to show the equivalent formula

$$\partial_t^\alpha f(t) = \frac{1}{\Gamma(1-\alpha)} \frac{f(t) - f(0)}{t^\alpha} + \frac{\alpha}{\Gamma(1-\alpha)} \int_0^t \frac{f(t) - f(\tau)}{(t-\tau)^{1+\alpha}} d\tau.$$

If in addition we set $f(t) = f(0)$ for $t < 0$, then we also get the following equivalent formulation

$$\partial_t^\alpha f(t) = \frac{\alpha}{\Gamma(1-\alpha)} \int_{-\infty}^t \frac{f(t) - f(\tau)}{(t-\tau)^{1+\alpha}} d\tau, \tag{2.1.4}$$

which is known as a (one-sided) nonlocal derivative in the sense of Marchaud [56, Section 5.4]. Clearly, both generalizations of the classical Caputo derivative have the advantage of working with "less than C^1-smooth" functions f (see, for instance, [46], for a more extensive comparison of these notions). Indeed, assuming C^1-regularity may place some severe restrictions on the problem [59]. Note that the integral (2.1.4) is well-defined for bounded functions that satisfy a local Hölder condition, $f \in C^{0,\lambda}((0,T);Y)$ with $\lambda > \alpha$ (also this may be weakened to $\lambda = \alpha$, if a bounded f belongs to a kind of Hölder space $H^{\alpha,-a}$, $a > 1$, that consists of functions satisfying a Hölder condition with a logarithmic "correction", see [56, Definition 1.7]). The integral definition in (2.1.2) seems to be more natural for the treatment of the nonlinear problem (see Chap. 3). Furthermore, (2.1.2) also appears as a more suitable notion to establish some compactness criteria and time regularity estimates for nonlinear time fractional PDEs; some examples involve time fractional compressible Navier–Stokes equations and time fractional Keller-Segel equations (see [45]). Finally, in the classical case when $\alpha = 1$, we let $\partial_t^1 := d/dt \; (= \partial_t)$.

Remark 2.1.2 It is worth mentioning the following facts.

(a) Firstly, we notice that if $f(0) = 0$, then (2.1.3) holds without the C^1-regularity assumption on f. Secondly, we mention that if the Banach space Y has the Radon-Nikodym property, then (2.1.3) holds for every function $f \in AC([0,T];Y)$ (that is, the space of all absolutely continuous functions on $[0,T]$ with values in Y). The latter space coincides with the classical vector-valued Sobolev space $W^{1,1}((0,T);Y)$. More emphasis on this topic will be given in Chap. 3 (after Remark 3.3.3).

(b) If one considers Eq. (1.0.1) with the classical Caputo fractional derivative, that is,

$$g_{1-\alpha} * \partial_t u = Au + f(x,t,u) \text{ in } \mathcal{X} \times (0,T], \; u(\cdot,0) = u_0 \text{ in } \mathcal{X}, \tag{2.1.5}$$

then Eqs. (1.0.1) and (2.1.5) will have the same mild solutions given by the integral representation in Definition 3.1.1. In addition, far away from $t = 0$, they will enjoy the same regularity. The main difference between solutions of the mentioned equations is their behavior when the time t is close to 0. We also mention that strong solutions of (1.0.1) and (2.1.5) coincide provided that some assumptions on the nonlinearity are satisfied (see Theorem 3.3.4).

For $0 < \alpha \leq 1$ we consider first the following nonhomogeneous linear Cauchy problem

$$\partial_t^\alpha u(t) = Au(t) + f(t), \quad t \in (0, T), \quad u(0) = u_0. \tag{2.1.6}$$

Here, A is a closed linear operator with domain $D(A)$ in a Banach space Y, $f : [0, \infty) \to Y$ is a given function and u_0 is a given vector in Y. The system (2.1.6) reduces to the classical parabolic equation when $\alpha = 1$. Our main goal is to investigate the problem of well-posedness for (2.1.6) in an unifying fashion, in all cases $0 < \alpha \leq 1$; this requires us to eventually introduce a suitable representation of solutions for (2.1.6). This will be done by means of the mild solution theory using integral solutions which we investigate subsequently. Thus, we need first to discuss the operators involved in such integral solutions.

Firstly, recall the definition of the Wright type function [37, Formula (28)] (see also [52, 56, 67]) is

$$\Phi_\alpha(z) := \sum_{n=0}^{\infty} \frac{(-z)^n}{n!\Gamma(-\alpha n + 1 - \alpha)}, \quad 0 < \alpha < 1, \quad z \in \mathbb{C}. \tag{2.1.7}$$

This is also sometimes called the Mainardi function. Following [14, p.14] (see also [37]), $\Phi_\alpha(t)$ is a probability density function, namely,

$$\Phi_\alpha(t) \geq 0, \quad t > 0; \quad \int_0^\infty \Phi_\alpha(t)dt = 1.$$

Furthermore, $\Phi_\alpha(0) = 1/\Gamma(1-\alpha)$ and the following formula on the moments of the Wright function is well-known (see [37]),

$$\int_0^\infty t^p \Phi_\alpha(t)dt = \frac{\Gamma(p+1)}{\Gamma(\alpha p + 1)}, \quad p > -1, \quad 0 < \alpha < 1. \tag{2.1.8}$$

For more details on the Wright type functions, we refer the reader to [14, 37, 48, 67] and the references therein. We note that the Wright functions have been used by Bochner to construct fractional powers of semigroup generators [68, Chapter IX].

Secondly, assume that the operator A generates a strongly continuous semigroup $S = (S(t))_{t \geq 0}$ on Y. For $t > 0$ we define the two operators

$$S_\alpha(t) : Y \to Y, \quad P_\alpha(t) : Y \to Y$$

as follows. For $v \in Y$, we set

$$\begin{cases} S_\alpha(t)v := \int_0^\infty \Phi_\alpha(\tau)S(\tau t^\alpha)v d\tau, \\ \\ P_\alpha(t)v := \alpha t^{\alpha-1} \int_0^\infty \tau\Phi_\alpha(\tau)S(\tau t^\alpha)v d\tau. \end{cases} \tag{2.1.9}$$

The operators S_α and P_α are known in the literature as resolvent families. In addition one does not need that A generates a semigroup in order to define resolvent families associated with A. The notion of resolvent families, their fine properties and their relations to abstract Cauchy and/or Volterra kind of equations have been introduced and intensively studied by Prüss in the monograph [53]. These notions have been used and extended by several authors. We refer for example to [4, 14, 47] and their references. The representation of resolvent families given in (2.1.9) also known as the principle of subordination has been introduced and studied for example in [14, 42–44] and their references. Finally, both formulas for the operators S_α, P_α in (2.1.9) can be recasted in terms of Mittag-Leffler functions (see, for instance, [28, Theorem 4.2]), but this further connection is not essential in our subsequent analysis.

We note that the semigroup property **does not** hold for the operators S_α, P_α, namely, $S_\alpha(t+s) \neq S_\alpha(t)\, S_\alpha(s)$, for all $t, s \geq 0$ (and the same is valid for P_α) unless $\alpha = 1$. Moreover, by definition the operator S_α is **strongly continuous**, that is,

$$\lim_{t\to 0^+} \|S_\alpha(t)\, v - v\|_Y = 0 \tag{2.1.10}$$

provided that $v \in Y$ but this is not the case for P_α unless $\alpha = 1$. Namely for $0 < \alpha < 1$, we have that $\lim_{t\to 0^+} \|t^{1-\alpha} P_\alpha(t)v - \Gamma(\alpha)v\|_Y = 0$ for every $v \in Y$. With these definitions, we have the following properties.

Proposition 2.1.3 *Let A with domain $D(A)$ be a closed and linear operator on the Banach space Y. Assume that A generates a strongly continuous and bounded semigroup $(S(t))_{t\geq 0}$ on Y, that is, there exists a constant $M > 0$ such that for all $t \geq 0$ and $f \in Y$, we have*

$$\|S(t)f\|_Y \leq M\|f\|_Y.$$

Then there exists a constant $C > 0$ such that for all $t \geq 0$ and $f \in Y$ we have

$$\|S_\alpha(t)f\|_Y \leq C\|f\|_Y \text{ and } \|t^{1-\alpha} P_\alpha(t)f\|_Y \leq C\|f\|_Y. \tag{2.1.11}$$

Proof The proposition follows by a simple application of the representation (2.1.9). □

Finally, in the remainder part of this subsection, we consider the solvability of the linear problem (2.1.6). We use the following notion of smooth solution for (2.1.6).

Definition 2.1.4 Let $0 < T_1 \leq T_2 < T$. We say that u is a strong solution for (2.1.6) on the interval $I = [0, T]$, if the following conditions are satisfied:

(i) (***The case*** $\alpha = 1$). The function $u \in C([0, T); Y)$ such that $u(0) = u_0$, $u(t) \in D(A)$, for all $t \in [T_1, T_2] \subset I$, and $\partial_t u \in C([T_1, T_2]; Y)$. Moreover, the differential equation $\partial_t u(t) = Au(t) + f(t)$ is satisfied on $[T_1, T_2] \subset I$.
(ii) (***The case*** $\alpha \in (0, 1)$). The function $u \in C([0, T), Y)$ such that $u(0) = u_0$, $u(t) \in D(A)$ for $t \in [T_1, T_2]$, and

$$\partial_t^\alpha u = \frac{d}{dt}\Big(g_{1-\alpha} * (u - u(0))\Big) \in C([T_1, T_2]; Y). \tag{2.1.12}$$

The differential equation $\partial_t^\alpha u(t) = Au(t) + f(t)$ is satisfied on $[T_1, T_2]$.

Remark 2.1.5 Let $\alpha \in (0, 1)$. Observe that (2.1.12) implies that $u \in C^{0,\alpha}([T_1, T_2]; Y)$ but not vice-versa, see, for instance, [24]. To the best of our knowledge, our notion of strong solutions has not been studied in the literature. For fractional in time Cauchy problems when $\alpha \in (0, 1)$, the notion of mild and classical solutions have been intensively studied. In this case, several authors have also investigated the asymptotic behavior of mild and classical solutions of (2.1.6). We refer the reader to the following references [4, 14, 25, 36, 42–44, 60] for further details. Our notion of strong solutions does not enjoy all the required regularity properties of classical solutions. For problem (2.1.6), a comparable notion of strong solutions that is similar to our kind has only been investigated in the reference [60] under the assumption that the operator A is quasi-sectorial on a Banach space Y, which also guarantees that A generates an analytic semigroup on Y. However, we emphasize that the linear problem (2.1.6) is not the only concern of the present paper. This constitutes only an appetizer that aims to give us some useful tools to investigate the semi-linear problem which is our main objective in subsequent chapters.

The existence of a strong solution in the sense of Definition 2.1.4, for problem (2.1.6) in the case $\alpha = 1$, is well known when A is a sectorial operator in Y and f is a suitable Hölder-continuous function (cf. [41, Theorem 3.2.2]). In what follows, we aim to deduce the corresponding result in the case $\alpha \in (0, 1)$ under similar conditions on A and f. Since the corresponding operators S_α, P_α are no longer semigroups when $\alpha \in (0, 1)$, the existence result requires a non-trivial proof; this extends the classical result ($\alpha = 1$) due to Henry [41, Theorem 3.2.2] to the non-standard case $\alpha \in (0, 1)$. But first, we need the following result.

Lemma 2.1.6 *Let $\alpha \in (0,1)$ and $q \in (1/\alpha, \infty)$ such that $\int_0^{T_0} \|f(t)\|_Y^q dt < \infty$, for some $T_0 > 0$. In addition, assume that A is sectorial in Y and $f \in C^{0,\beta}((0,T);Y)$ for some $\beta > 0$. For $0 \le t < T$, we define*

$$F(t) := \int_0^t P_\alpha(t-\tau) f(\tau) ds. \tag{2.1.13}$$

Then $F \in C([0,T);Y) \cap C^{0,\gamma}((0,T);Y)$, for some $\gamma > 0$ and $F(0) = 0$. In addition, we have $(g_{1-\alpha} * (F - F(0)) \in C^1((0,T);Y)$, $F(t) \in D(A)$ for $0 < t < T$ and

$$\partial_t^\alpha F(t) = AF(t) + f(t), \quad 0 < t < T.$$

Proof We prove the lemma in several steps.

Step 1 First we claim that $F \in C([0,T);Y)$ and $F(0) = 0$. Define for a sufficiently small real number $\rho > 0$,

$$F_\rho(t) := \begin{cases} \int_0^{t-\rho} P_\alpha(t-s) f(s) ds, & \text{for } t > \rho \\ 0, & \text{for } t \le \rho \end{cases}$$

(and also set $f(s) = 0$, for $s < 0$ and $s > T$). We then notice that there exists a constant $C_\alpha > 0$ depending only on α such that (by using (2.1.11))

$$\begin{aligned} \left\| F_\rho(t) - F(t) \right\|_Y &\le C_\alpha \int_{t-\rho}^t \|P_\alpha(t-\tau)\|_Y \|f(\tau)\|_Y d\tau \qquad (2.1.14) \\ &\le C_\alpha \int_{t-\rho}^t (t-\tau)^{\alpha-1} \|f(\tau)\|_Y d\tau \\ &\le C_\alpha \left(\int_{t-\rho}^t (t-\tau)^{p(\alpha-1)} d\tau \right)^{1/p} \left(\int_{t-\rho}^t \|f(\tau)\|_Y^q d\tau \right)^{1/q}, \end{aligned}$$

where $p(1 - 1/q) = 1$ and we have used the Hölder inequality for the last estimate. Since $q > \frac{1}{\alpha}$ and hence, $p < \frac{1}{1-\alpha}$, clearly $\varepsilon_\alpha := p(\alpha - 1) + 1 > 0$. Obviously,

$$\int_{t-\rho}^t (t-\tau)^{p(\alpha-1)} d\tau = \frac{\rho^{\varepsilon_\alpha}}{\varepsilon_\alpha} \to 0 \text{ as } \rho \to 0,$$

so that the right-hand side of (2.1.14) tends to zero as $\rho \to 0^+$, uniformly in $0 \le t \le T_0$ with $T_0 < T$. Namely, F is the uniform limit as $\rho \to 0$ of the sequence F_ρ.

Analogously, for $t \in (0, T_0)$, we have

$$\begin{aligned}
\|F(t)\|_Y &\le \int_0^t \|P_\alpha(t-\tau)\|_Y \, \|f(\tau)\|_Y \, d\tau \\
&\le C_\alpha \left(\int_0^t (t-\tau)^{p(\alpha-1)} \, d\tau \right)^{1/p} \left(\int_0^t \|f(\tau)\|_Y^q \, d\tau \right)^{1/q} \\
&\le C_{\alpha,p} t^{\varepsilon_\alpha} \left(\int_0^{T_0} \|f(\tau)\|_Y^q \, d\tau \right)^{1/q},
\end{aligned}$$

which tends to zero as $t \to 0^+$; the latter estimate proves the continuity of F at $t = 0$. To show that F is continuous on $[0, T)$, it suffices to show that F_ρ is continuous on $[0, T)$ since F is the uniform limit (as $\rho \to 0$) of F_ρ. We thus need to look at the difference

$$\begin{aligned}
&F_\rho(t+h) - F_\rho(t) \qquad (2.1.15) \\
&= \int_0^{t-\rho} \left(P_\alpha(t+h-\tau) - P_\alpha(t-\tau)\right) f(\tau) d\tau + \int_{t-\rho}^{t+h-\rho} P_\alpha(t+h-\tau) f(\tau) \, d\tau,
\end{aligned}$$

for $0 < t \le t+h \le T_0 \le T$. In a similar fashion, we can estimate the second summand in (2.1.15), for any $\rho > 0$,

$$\begin{aligned}
&\left\| \int_{t-\rho}^{t+h-\rho} P_\alpha(t+h-\tau) f(\tau) \, ds \right\|_Y \\
&\le C_\alpha \left(\int_{t-\rho}^{t+h-\rho} \|f(\tau)\|_Y^q \, d\tau \right)^{1/q} \left(\int_{t-\rho}^{t+h-\rho} (t+h-\tau)^{p(\alpha-1)} d\tau \right)^{1/p} \\
&\le C_{\alpha,p} \Big((\rho+h)^{\varepsilon_\alpha} - \rho^{\varepsilon_\alpha} \Big) \left(\int_{t-\rho}^{t+h-\rho} \|f(\tau)\|_Y^q \, d\tau \right)^{1/q}
\end{aligned}$$

which once again tends to zero as $h \to 0$ since $\varepsilon_\alpha > 0$. For the first summand in (2.1.15), it is clear that

$$\lim_{h \to 0} \| \left(P_\alpha(t+h-\tau) - P_\alpha(t-\tau)\right) f(s) \|_Y = 0. \qquad (2.1.16)$$

In addition, by assumption (indeed, since A is sectorial, we infer from (2.1.9) that P_α is analytic for $t > 0$, and $t^{1-\alpha} \|P_\alpha(t)\|_{\mathcal{L}(Y,Y)} \le C_\alpha$), the following estimate holds:

$$\begin{aligned}
\| \left(P_\alpha(t+h-\tau) - P_\alpha(t-\tau)\right) f(\tau) \|_Y &\le \|P_\alpha(t+h-\tau) f(\tau)\|_Y + \|P_\alpha(t-\tau) f(\tau)\|_Y \\
&\le C_\alpha \left((t+h-\tau)^{\alpha-1} + (t-\tau)^{\alpha-1} \right) \|f(\tau)\|_Y \\
&\le 2C_\alpha (t-\tau)^{\alpha-1} \|f(\tau)\|_Y. \qquad (2.1.17)
\end{aligned}$$

Since

$$\begin{aligned}\int_0^{t-\rho}(t-\tau)^{\alpha-1}\|f(\tau)\|_Y d\tau &\le \left(\int_0^{t-\rho}(t-\tau)^{p(\alpha-1)}d\tau\right)^{1/p}\left(\int_0^{t-\rho}\|f(\tau)\|_Y^q d\tau\right)^{1/q}\\ &\le C_{\alpha,p}(t-\rho)^{\frac{\varepsilon_\alpha}{p}}\left(\int_0^{t-\rho}\|f(\tau)\|_Y^q d\tau\right)^{1/q}\\ &< \infty,\end{aligned} \tag{2.1.18}$$

it follows from the Lebesgue dominated convergence theorem (by using (2.1.16)–(2.1.18)) that

$$\int_0^{t-\rho}\|(P_\alpha(t+h-\tau)-P_\alpha(t-\tau))f(\tau)\|_Y d\tau \to 0 \text{ as } h\to 0.$$

We have shown that $F_\rho \in C([0,T);Y)$ and the claim that $F \in C([0,T);Y)$ is immediate. Next, we check that $F \in C^{0,\gamma}((0,T);Y)$ for some $0<\gamma<1$. Indeed, let $0<t_1<t_2<T$ and observe that

$$\begin{aligned}F(t_2)-F(t_1) &= \int_0^{t_1} P_\alpha(\tau)(f(t_2-\tau)-f(t_1-\tau))\,d\tau + \int_{t_1}^{t_2} P_\alpha(\tau)f(t_2-\tau)d\tau\\ &= \int_0^{t_1} P_\alpha(\tau)(f(t_2-\tau)-f(t_1-\tau))\,d\tau + \int_0^{t_2-t_1} P_\alpha(t_2-\tau)f(\tau)d\tau.\end{aligned}$$

For the second summand, using the Hölder inequality we get that there exists a constant $C = C(\alpha, p) > 0$ such that

$$\begin{aligned}\int_0^{t_2-t_1}\|P_\alpha(t_2-\tau)f(\tau)\|_Y d\tau \le& C_\alpha\int_0^{t_2-t_1}(t_2-\tau)^{\alpha-1}\|f(\tau)\|_Y d\tau\\ \le& C_\alpha\left(\int_0^{t_2-t_1}(t_2-\tau)^{p(\alpha-1)}\,d\tau\right)^{\frac{1}{p}}\left(\int_0^{t_2-t_1}\|f(\tau)\|_Y^q d\tau\right)^{\frac{1}{q}}\\ \le& C_{\alpha,p}\left(t_2^{\varepsilon_\alpha}-t_1^{\varepsilon_\alpha}\right)^{\frac{1}{p}}\left(\int_0^{t_2-t_1}\|f(\tau)\|_Y^q d\tau\right)^{\frac{1}{q}}.\end{aligned}$$

Using the mean value theorem and the fact that the mapping $t \mapsto t^{p(\alpha-1)}$ is decreasing on $[t_1, t_2]$, we get that

$$t_2^{\varepsilon_\alpha} - t_1^{\varepsilon_\alpha} \le \varepsilon_\alpha t_1^{p(\alpha-1)}(t_2-t_1).$$

This together with the previous estimate give

$$\int_0^{t_2-t_1}\|P_\alpha(t_2-\tau)f(\tau)\|_Y d\tau \le C_{\alpha,p}t_1^{\alpha-1}|t_2-t_1|^{\frac{1}{p}}\left(\int_0^{t_2-t_1}\|f(\tau)\|_Y^q d\tau\right)^{\frac{1}{q}}. \tag{2.1.19}$$

For the first summand, we use the fact that $f \in C^{0,\beta}((0,T);Y)$, to estimate

$$\begin{aligned}
&\int_0^{t_1} \|P_\alpha(\tau)\,(f(t_2-\tau)-f(t_1-\tau))\,\|_Y d\tau \qquad (2.1.20)\\
&\le C_\alpha \int_0^{t_1} \tau^{\alpha-1}\|f(t_2-\tau)-f(t_1-\tau)\|_Y d\tau\\
&\le C_\alpha M_f |t_2-t_1|^\beta \int_0^{t_1} \tau^{\alpha-1}\, d\tau\\
&\le C_\alpha M_f t_1^\alpha |t_2-t_1|^\beta,
\end{aligned}$$

where M_f denotes the Hölder constant of f. Letting $\gamma := \min\{1/p,\beta\}$, it follows from (2.1.19) and (2.1.20) that $\|F(t_2)-F(t_1)\|_Y = O(|t_2-t_1|^\gamma)$. Hence, $F \in C^{0,\gamma}((0,T);Y)$.

Step 2 We show that $F(t) \in D(A)$ for all $0 < t < T$. We use the fact that A is a sectorial operator in Y, namely, $\|AS(t)\|_{\mathcal{L}(Y)} \le Ct^{-1}$, for $t \in (0,T)$. We can exploit the formula (2.1.9) for P_α, to find that $\left\|t^{1-\alpha} AP_\alpha(t)\right\|_{\mathcal{L}(Y)} \le C_\alpha t^{-\alpha}$, and so

$$\|AP_\alpha(t)\|_{\mathcal{L}(Y)} \le C_\alpha t^{-1}, \quad \text{for } t \in (0,T). \qquad (2.1.21)$$

In particular, this implies that as $0 \le \tau < t$ and $f(\tau) \in Y$, $P_\alpha(t-\tau)f(\tau) \in D(A)$, and the Riemann sums for $F_\rho(t)$ $(t > \rho)$, namely $\sum_{\tau_i \le t-\rho} P_\alpha\left(t-\tau_j\right) f\left(\tau_j\right) \Delta\tau_j$, also belongs to $D(A)$. Moreover,

$$\begin{aligned}
\lim_{\Delta\tau\to 0} A \sum_{\tau\le t-\rho} P_\alpha(t-\tau)\,f(\tau)\,\Delta\tau &= \lim_{\Delta\tau\to 0} \sum_{\tau\le t-\rho} AP_\alpha(t-\tau)\,f(\tau)\,\Delta\tau\\
&= \int_0^{t-\rho} AP_\alpha(t-\tau)f(\tau)d\tau.
\end{aligned}$$

This yields $F_\rho(t) \in D(A)$ since A is closed. On the other hand, we have

$$\begin{aligned}
AF_\rho(t) &= \int_0^{t-\rho} AP_\alpha(t-\tau)f(\tau)d\tau \qquad (2.1.22)\\
&= \int_0^{t-\rho} AP_\alpha(t-\tau)\,(f(\tau)-f(t))\,d\tau + (S_\alpha(\rho)-S_\alpha(t))\,f(t).
\end{aligned}$$

Thus owing to (2.1.22), as $\rho \to 0^+$, it holds

$$AF_\rho(t) \to \int_0^t AP_\alpha(t-\tau)\,(f(\tau)-f(t))\,d\tau + (I-S_\alpha(t))\,f(t). \qquad (2.1.23)$$

Indeed, the second summand in (2.1.23) follows from (2.1.22) in view of the continuity of $S_\alpha(\rho)$ as $\rho \to 0^+$. For the first summand, it is a consequence of the following estimate and (2.1.21). More precisely, using by assumption that $\|f(\tau) - f(t)\|_Y = O(|t-\tau|^\beta)$, for some $\beta > 0$,

$$\left\| \int_{t-\rho}^{t} AP_\alpha(t-\tau)(f(\tau) - f(t))\,d\tau \right\|_Y \overset{(2.1.21)}{\leq} C_\alpha \int_{t-\rho}^{t} (t-\tau)^{\beta-1}\,d\tau = C_{\alpha,\beta}\rho^\beta,$$

and this tends to zero as $\rho \to 0^+$. Hence, the claim in (2.1.23) is immediate; since A is closed, $AF_\rho(t) \to AF(t)$ as $\rho \to 0^+$, from which we infer that $F(t) \in D(A)$ for $0 < t < T$.

Step 3 We finally claim that $g_{1-\alpha} * (F - F(0)) \in C^1((0,T);Y)$ and F satisfies a proper differential equation. Since $F(0) = 0$, we have

$$(g_{1-\alpha} * (F - F(0)))(t) = (g_{1-\alpha} * F)(t) = (g_{1-\alpha} * P_\alpha * f)(t) = (S_\alpha * f)(t).$$

The operator $S_\alpha(t)$ is analytic for $t > 0$, so that the mapping $s \mapsto (S_\alpha * f)(t)$ is differentiable and

$$\frac{d}{dt}\Big[(S_\alpha * f)(t)\Big] = \frac{d}{dt}\int_0^t S_\alpha(t-\tau)f(\tau)d\tau = S_\alpha(0)f(t) + \int_0^t S_\alpha'(t-\tau)f(\tau)d\tau \tag{2.1.24}$$

$$= f(t) + \int_0^t AP_\alpha(t-\tau)f(\tau)d\tau = f(t) + A\int_0^t P_\alpha(t-\tau)f(\tau)d\tau.$$

This also shows in particular that F is Riemann-Liouville (as well as Caputo) differentiable of order α and $D_t^\alpha F(t) = \partial_t^\alpha F(t)$ for all $0 < t < T$. Since A is sectorial, $P_\alpha(t)u, S_\alpha(t)u \in D(A)$ for every $u \in Y$ and for every $t > 0$. In addition we have that the mapping $t \mapsto S_\alpha(t)u$ is differentiable with $\frac{d}{dt}\Big(S_\alpha(t)u\Big) = AP_\alpha(t)u$. Let now $G(t) := \int_0^t S_\alpha(t-\tau)f(\tau)d\tau$. Then by (2.1.24), G is differentiable with $G'(t) = f(t) + AF(t)$. On the other hand, we clearly have $G(t) = (g_{1-\alpha} * P_\alpha * f)(t)$. This implies that

$$G'(t) = \frac{d}{dt}\Big(g_{1-\alpha} * P_\alpha * f\Big)(t) = D_t^\alpha(P_\alpha * f)(t) = D_t^\alpha F(t) = \partial_t^\alpha F(t),$$

in view of the fact that $F(0) = 0$. We have shown that

$$\partial_t^\alpha F(t) = AF(t) + f(t),\ 0 < t < T.$$

The proof of lemma is finished. □

We can then conclude with the following existence result for problem (2.1.6).

Theorem 2.1.7 *Assume A is sectorial in Y and let $u_0 \in Y$. Consider the following two cases.*

(a) **(The case** $\alpha = 1$*). Let $f \in C^{0,\beta}((0,T);Y)$, $\beta > 0$ and assume $\int_0^{T_0} \|f(t)\|_Y dt < \infty$, for some $T_0 > 0$.*
(b) **(The case** $\alpha \in (0,1)$*). Let $q \in (1/\alpha, \infty)$ such that $\int_0^{T_0} \|f(t)\|_Y^q dt < \infty$ for some $T_0 > 0$ and $f \in C^{0,\beta}((0,T);Y)$, for some $\beta > 0$.*

Then there exists a unique strong solution of problem (2.1.6) in the sense of Definition 2.1.4. This strong solution is given by

$$u(t) = S(t)u_0 + \int_0^t S(t-\tau)f(\tau)d\tau$$

in the first case (a), and by

$$u(t) = S_\alpha(t)u_0 + \int_0^t P_\alpha(t-\tau)f(\tau)d\tau$$

in the second case (b), respectively.

Proof Case (a) is proved in [41, Theorem 3.2.2]. We complete the proof of the second case exploiting the conclusions of Lemma 2.1.6. Let $v(t) := S_\alpha(t)u_0 = (g_{1-\alpha} * P_\alpha)(t)u_0$. Since A generates an analytic semigroup on Y, the mapping $t \mapsto S_\alpha(t)u_0$ is differentiable for any $t > 0$. In addition, for all $0 < t < T$,

$$\begin{aligned} \partial_t^\alpha v(t) = & \int_0^t g_{1-\alpha}(t-\tau)S_\alpha'(\tau)u_0 d\tau = \int_0^t g_{1-\alpha}(t-\tau)AP_\alpha(\tau)u_0 d\tau \qquad (2.1.25) \\ =& A\int_0^t g_{1-\alpha}(t-\tau)P_\alpha(\tau)u_0 d\tau = Av(t). \end{aligned}$$

Let $w(t) := \int_0^t P_\alpha(t-\tau)f(\tau)d\tau$. Then Lemma 2.1.6 yields $\partial_t^\alpha w(t) = Aw(t) + f(t)$, for all $0 < t < T$. Clearly, $u = v + w$ is the desired solution of (2.1.6). Uniqueness is easy to see and the proof is finished. □

Remark 2.1.8 In the second statement of Theorem 2.1.7 when $\alpha \in (0,1)$, the assumption that A generates an analytic semigroup on Y is actually not required. This is only necessary for the case $\alpha = 1$ as treated in the monograph [41]. In the case $\alpha \in (0,1)$, it has been shown in [14, Theorem 3.4] that if A generates a strongly continuous semigroup (that is not necessarily analytic in Y), then the operators $S_\alpha(t)$ and $P_\alpha(t)$ are automatically analytic for every $t > 0$. Notice that these properties were the only required ingredients in the proof of the second alternative (b).

One has additional smoothness for the strong solution of (2.1.6) if f is smooth enough. The result can prove quite useful in those instances when the multiplication of (2.1.6) by the test function $\partial_t u$ is required in order to perform additional estimates for the corresponding solutions of (2.1.6).

Proposition 2.1.9 *Let $f \in W^{1,q}((0,T);Y)$ for some $q \in (1/\alpha, \infty]$ and assume that A generates a strongly continuous semigroup. Let u be a strong solution in the sense of Definition 2.1.4 in the case when $\alpha \in (0,1)$. Then it also holds that $u \in C^1((0,T);Y)$.*

Proof This follows directly by exploiting the formula for the integral solution owing to the fact that

$$u'(t) = S_\alpha'(t)u_0 + P_\alpha(t)f(0) + \int_0^t P_\alpha(t-\tau)f'(\tau)d\tau. \tag{2.1.26}$$

In view of Remark 2.1.8, the families $S_\alpha(t)$ and $P_\alpha(t)$ are both analytic as functions of $t > 0$; in particular, $S_\alpha'(t) = AP_\alpha(t)$, for all $t > 0$. Obviously since $f \in W^{1,q}((0,T);Y)$, the value $f(0) \in Y$ is well-defined. Both the first two summands in (2.1.26) belong to $C((0,T);Y)$; for the last summand, one argues exactly as in the proof of Lemma 2.1.6 to show that $P_\alpha * f' \in C([0,T);Y)$. The proof is finished. □

2.2 Ultracontractivity and Resolvent Families

In this subsection we assume that our Banach space $Y = L^2(\mathcal{X})$ and that the operator A generates a strongly continuous semigroup $(S(t))_{t\geq 0}$ on $L^2(\mathcal{X})$.

Proposition 2.2.1 *Assume that the semigroup S is submarkovian in the sense that it is positive and L^∞-contractive, that is,*

$$S(t)u \geq 0 \text{ for all } t \geq 0, \text{ whenever } 0 \leq u \in L^2(\mathcal{X}),$$

and

$$\|S(t)f\|_{L^p(\mathcal{X})} \leq \|f\|_{L^p(\mathcal{X})} \text{ for all } f \in L^p(\mathcal{X}) \cap L^2(\mathcal{X}),\ p \in [1,\infty]. \tag{2.2.1}$$

Then $S_\alpha(t)$ and $P_\alpha(t)$ are positive and for all $f \in L^p(\mathcal{X}) \cap L^2(\mathcal{X})$, $p \in [1,\infty]$, we have

$$\|S_\alpha(t)f\|_{L^p(\mathcal{X})} \leq \|f\|_{L^p(\mathcal{X})} \text{ and } \|\Gamma(\alpha)t^{1-\alpha}P_\alpha(t)f\|_{L^p(\mathcal{X})} \leq \|f\|_{L^p(\mathcal{X})}. \tag{2.2.2}$$

Proof Since $S(t)$ is a positive operator, it follows from (2.1.9) that $S_\alpha(t)$ and $P_\alpha(t)$ are also positive. Let $f \in L^p(\mathcal{X}) \cap L^2(\mathcal{X})$ for $p \in [1,\infty]$. Since $\|S(t)\|_{\mathcal{L}(L^p(\mathcal{X}))} \leq 1$

for all $t \geq 0$, it follows from (2.1.9) that

$$\begin{aligned}\|S_\alpha(t)f\|_{L^p(\mathcal{X})} &\leq \int_0^\infty \Phi_\alpha(\tau)\|S(\tau t^\alpha)f\|_{L^p(\mathcal{X})}d\tau \\ &\leq \int_0^\infty \Phi_\alpha(\tau)d\tau\|f\|_{L^p(\mathcal{X})} = \|f\|_{L^p(\mathcal{X})}.\end{aligned}$$

Similarly it follows from (2.1.9) and (2.1.8) that

$$\begin{aligned}\|\Gamma(\alpha)t^{1-\alpha}P_\alpha(t)f\|_{L^p(\mathcal{X})} \leq& \alpha\Gamma(\alpha)t^{\alpha-1}t^{1-\alpha}\int_0^\infty \tau\Phi_\alpha(\tau)d\tau\|f\|_{L^p(\mathcal{X})} \\ =&\alpha\Gamma(\alpha)\frac{\Gamma(2)}{\Gamma(\alpha+1)}\|f\|_{L^p(\mathcal{X})} = \|f\|_{L^p(\mathcal{X})}.\end{aligned}$$

We have shown (2.2.2) and the proof is finished. □

Proposition 2.2.2 *Assume that the semigroup S is ultracontractive in the sense that for every $1 \leq p \leq q \leq \infty$, the operator $S(t)$ maps $L^p(\mathcal{X})$ into $L^q(\mathcal{X})$ and there exist two constants $C > 0$, $\beta_A > 0$ such that for all $t > 0$,*

$$\|S(t)\|_{\mathcal{L}(L^p(\mathcal{X}),L^q(\mathcal{X}))} \leq Ct^{-\beta_A\left(\frac{1}{p}-\frac{1}{q}\right)}. \tag{2.2.3}$$

Here the constant β_A is assumed independent of p, q. Then the following assertions hold.

(a) If $\beta_A\left(p^{-1}-q^{-1}\right) < 1$, then there exists $C > 0$ such that for all $t > 0$,

$$\|S_\alpha(t)\|_{\mathcal{L}(L^p(\mathcal{X}),L^q(\mathcal{X}))} \leq Ct^{-\alpha\beta_A\left(\frac{1}{p}-\frac{1}{q}\right)}. \tag{2.2.4}$$

(b) If $\beta_A\left(p^{-1}-q^{-1}\right) < 2$, then there exists $C > 0$ such that for all $t > 0$,

$$\|t^{1-\alpha}P_\alpha(t)\|_{\mathcal{L}(L^p(\mathcal{X}),L^q(\mathcal{X}))} \leq Ct^{-\alpha\beta_A\left(\frac{1}{p}-\frac{1}{q}\right)}. \tag{2.2.5}$$

Proof Let $1 \leq p \leq q \leq \infty$ and $f \in L^p(\mathcal{X})$.

(a) Assume that $\beta_A\left(p^{-1}-q^{-1}\right) < 1$. Using (2.1.9), (2.2.3) and (2.1.8) we deduce that

$$\begin{aligned}\|S_\alpha(t)f\|_{L^q(\mathcal{X})} &\leq \int_0^\infty \Phi_\alpha(\tau)\|S(\tau t^\alpha)f\|_{L^q(\mathcal{X})}d\tau \\ &\leq C\int_0^\infty \Phi_\alpha(\tau)\tau^{-\beta_A\left(\frac{1}{p}-\frac{1}{q}\right)}t^{-\alpha\beta_A\left(\frac{1}{p}-\frac{1}{q}\right)}\|f\|_{L^p(\mathcal{X})}d\tau\end{aligned}$$

$$\leq \|f\|_{L^p(\mathcal{X})} C t^{-\alpha\beta_A\left(\frac{1}{p}-\frac{1}{q}\right)} \int_0^\infty \Phi_\alpha(\tau)\tau^{-\beta_A\left(\frac{1}{p}-\frac{1}{q}\right)} d\tau$$

$$= C\frac{\Gamma(1-\beta_A\left(p^{-1}-q^{-1}\right))}{\Gamma(1-\alpha\beta_A\left(p^{-1}-q^{-1}\right))}\|f\|_{L^p(\mathcal{X})} t^{-\alpha\beta_A\left(\frac{1}{p}-\frac{1}{q}\right)}.$$

(b) Now assume that $\beta_A\left(p^{-1}-q^{-1}\right) < 2$. Then using (2.1.9), (2.2.3) and (2.1.8) we obtain

$$\|t^{1-\alpha} P_\alpha(t) f\|_{L^q(\mathcal{X})} \leq \alpha \int_0^\infty \tau\Phi_\alpha(\tau)\|S(\tau t^\alpha) f\|_{L^q(\mathcal{X})} d\tau$$

$$\leq \alpha C \int_0^\infty \Phi_\alpha(\tau)\tau^{1-\beta_A\left(\frac{1}{p}-\frac{1}{q}\right)} t^{-\alpha\beta_A\left(\frac{1}{p}-\frac{1}{q}\right)} \|f\|_{L^p(\mathcal{X})} d\tau$$

$$\leq \|f\|_{L^p(\mathcal{X})} \alpha C t^{-\alpha\beta_A\left(\frac{1}{p}-\frac{1}{q}\right)} \int_0^\infty \Phi_\alpha(\tau)\tau^{1-\beta_A\left(\frac{1}{p}-\frac{1}{q}\right)} d\tau$$

$$= \frac{\alpha\Gamma(2-\beta_A\left(p^{-1}-q^{-1}\right))}{\Gamma(1+\alpha-\alpha\beta_A\left(p^{-1}-q^{-1}\right))}\left(C\|f\|_{L^p(\mathcal{X})}\right) t^{-\alpha\beta_A\left(\frac{1}{p}-\frac{1}{q}\right)}.$$

The proof is finished. □

Remark 2.2.3 The explicit constant in (2.1.8) allows us to see that all the constants $C = C(\alpha) > 0$ involved in the estimates of Propositions 2.2.1, 2.2.2 are bounded as $\alpha \to 1$. In particular, all constants involved in subsequent estimates are also bounded as $\alpha \to 1$. In particular, these features allow us to recover the results in the case $\alpha = 1$ in a natural way.

Next, let V and H be Hilbert spaces such that $V \overset{d}{\hookrightarrow} H$. Recall that in that case we have the Gelfand triple

$$V \overset{d}{\hookrightarrow} H \overset{d}{\hookrightarrow} V^\star.$$

Let $\mathcal{E} : V \times V \to \mathbb{R}$ be a bilinear and symmetric form. Throughout the rest of the paper we will use the following terminology.

- We shall say that $\mathcal{E}$ is closed if $\{u_n\}_{n\in\mathbb{N}} \subset V$ is such that

$$\lim_{n,m\to\infty}\left[\mathcal{E}(u_n-u_m, u_n-u_m) + \|u_n-u_m\|_H^2\right] = 0,$$

then there exists $u \in V$ such that $\lim_{n\to\infty}\left[\mathcal{E}(u_n-u, u_n-u) + \|u_n-u\|_H\right] = 0$.

- We will say that it is continuous if there exists $M > 0$ such that

$$|\mathcal{E}(u,v)| \leq M\|u\|_V\|v\|_V \quad \text{for all } u, v \in V.$$

- We will say that $\mathcal{E}$ is H-elliptic, if there exist constant $C > 0$ and $\nu \in \mathbb{R}$ such that

$$\mathcal{E}(u, u) + \nu\|u\|_H^2 \geq C\|u\|_V^2, \quad \forall\, u \in V.$$

- Finally, $\mathcal{E}$ will be said coercive, if there exists $C > 0$ such that

$$\mathcal{E}(u, u) \geq C\|u\|_V^2 \ \text{ for all } \ u \in V.$$

It is clear that if $\mathcal{E}$ is coercive, then it is H-elliptic.

Remark 2.2.4 Now assume that $\mathcal{E}$ is closed, symmetric, continuous and H-elliptic. Define the operator

$$\mathcal{A} : V \to V^\star \ \text{ by } \langle \mathcal{A}u, v\rangle_{V^\star, V} := \mathcal{E}(u, v), \quad \forall\, u, v \in V. \tag{2.2.6}$$

Let $\mathbb{A}$ be the part of the operator $\mathcal{A}$ in H in the sense that

$$D(\mathbb{A}) = \{u \in V : \mathcal{A}u \in H\}, \quad \mathbb{A}u = \mathcal{A}u. \tag{2.2.7}$$

It is easy to show that

$$D(\mathbb{A}) = \Big\{u \in V : \exists\, w \in H \text{ and } \mathcal{E}(u, v) = (w, v)_H \ \forall\, v \in V\Big\}, \quad \mathbb{A}u = w. \tag{2.2.8}$$

In addition we have that the operator $A := -\mathbb{A}$ generates a strongly continuous and analytic semigroup $(e^{tA})_{t\geq 0}$ on H.

We give a general abstract setting that will imply ultracontractivity with the precise constant. A complete proof of this result and a more general version can be found in [51, Chapters 3, 4 and 6] (see also [27, Chapters 1 and 2]). Here and below we set $w \wedge 1 := \min\{1, w\}$ for any $w \geq 0$.

Theorem 2.2.5 *Let V be a Hilbert space such that $V \overset{d}{\hookrightarrow} L^2(\mathcal{X})$. Let $\mathcal{E}$ with domain V be a symmetric, closed, continuous and $L^2(\mathcal{X})$-elliptic bilinear form on $L^2(\mathcal{X})$. Let A be the self-adjoint operator on $L^2(\mathcal{X}, m)$ associated with $\mathcal{E}$ in the sense of Remark 2.2.4 and let $(e^{tA})_{t\geq 0}$ be the strongly continuous semigroup on $L^2(\mathcal{X})$ generated by A. Assume in addition that the semigroup is submarkovian in the sense that it is positive and $L^\infty(\mathcal{X})$-contractive. Then the following assertions are equivalent.*

(a) There exists a constant $C > 0$ such that for all $u \in \mathcal{V}$,

$$\|u\|^2_{L^{\frac{2\mu}{\mu-2}}(\mathcal{X})} \leq C\mathcal{E}(u, u) \ \text{ for some } \ \mu > 2. \tag{2.2.9}$$

(b) The semigroup is ultracontractive with constant $\beta_A = \frac{\mu}{2}$.

Remark 2.2.6 Let $V, \mathcal{E}$ and A be as in Theorem 2.2.5. The following situations are well known (see e.g. [27, 32, 51]).

(i) The semigroup $(e^{tA})_{t\geq 0}$ is positive if and only if for every $u \in V$ we have that $u^+ = \max\{u, 0\}$, $u^- = \max\{-u, 0\} \in V$ and

$$\mathcal{E}(u^+, u^-) \leq 0.$$

(ii) Assume that the semigroup $(e^{tA})_{t\geq 0}$ is positive. Then it is $L^\infty(\mathcal{X})$-contractive if and only if for every $0 \leq u \in V$, we have that $u \wedge 1 \in V$ and

$$\mathcal{E}(u \wedge 1, u \wedge 1) \leq \mathcal{E}(u, u).$$

In what follows, our main working assumption is that $(S(t))_{t\geq 0}$ is ***ultracontractive*** and ***submarkovian*** in the aforementioned sense. For instance, we note that the submarkovian property implies from [27, Theorem 1.4.1] that $(S(t))_{t\geq 0}$ can then be extended to contraction semigroups $S_p(t)$ on $L^p(\mathcal{X})$ for every $p \in [1, \infty]$, and each semigroup is strongly continuous if $p \in [1, \infty)$ and bounded analytic if $p \in (1, \infty)$. Denote by A_p the generator of the semigroup on $L^p(\mathcal{X})$ so that $A_2 \equiv A$. In that case if $p \in (2, \infty)$, then we have that A_p is the part of $A = A_2$ in $L^p(\mathcal{X}) \hookrightarrow L^2(\mathcal{X})$ in the sense that

$$D(A_p) = \left\{u \in D(A) \cap L^p(\mathcal{X}) :\ Au \in L^p(\mathcal{X})\right\},\ A_p u = Au. \tag{2.2.10}$$

The operator A_∞ is defined as $(\lambda - A_\infty)^{-1} = \left[(\lambda - A_1)^{-1}\right]^\star$ for all $\lambda > 0$. If A is an unbounded operator then the semigroup on $L^\infty(\mathcal{X})$ is not strongly continuous or equivalently $D(A_\infty)$ is not dense in $L^\infty(\mathcal{X})$. In any case we shall set

$$\mathcal{L}^\infty(\mathcal{X}) := \overline{D(A_\infty)}^{L^\infty(\mathcal{X})}.$$

We mention that in most situations we shall have

$$C(\overline{\mathcal{X}}) \subseteq \mathcal{L}^\infty(\mathcal{X}) \subset L^\infty(\mathcal{X}).$$

For more details on this topic we refer to the monograph [27, Chapter 1]. For $p \in (1, 2)$, a description of A_p exactly as in (2.2.10) is in general not an easy task. However, by [51, Theorem 3.9], for every $p \in (1, \infty)$, if $u \in D(A_p)$, then $u|u|^{\frac{p}{2}-1} \in D(\mathcal{E}) = V$ and

$$\begin{aligned}\frac{4(p-1)}{p^2}\mathcal{E}\left(u|u|^{\frac{p}{2}-1}, u|u|^{\frac{p}{2}-1}\right) \leq & \left(-A_p u, |u|^{p-1}\mathrm{sgn}u\right)_{L^2(\mathcal{X})} \\ \leq & 2\mathcal{E}\left(u|u|^{\frac{p}{2}-1}, u|u|^{\frac{p}{2}-1}\right).\end{aligned} \tag{2.2.11}$$

We also mention that the semigroups are also consistent in the sense that for all $t \geq 0$, $S_p(t) f = S_q(t) f$, for all $t \geq 0$ whenever $f \in L^p(\mathcal{X}) \cap L^q(\mathcal{X})$ and $A_q \subseteq A_p$ for any $q \geq p$; namely, $D(A_q) \subseteq D(A_p)$ and $A_q f = A_p f$, for all $f \in D(A_q)$. Since $L^p(\mathcal{X}) \subseteq L^1(\mathcal{X})$ (recall that $\mathcal{X}$ is a relatively compact Hausdorff space) and $S_p(t) \subseteq S_1(t)$ for all $p \in [1, \infty]$, we can drop the index p and merely write S for the semigroup, for the sake of notational simplicity. We shall also apply this convention to the operators S_α, P_α from (2.1.9). Clearly by (2.2.3), $S(t)$ defines a bounded (linear) operator from $L^p(\mathcal{X})$ into $L^q(\mathcal{X})$ for $1 \leq p \leq q \leq \infty$. For the sake of brevity, in what follows we may write (and define) its operator norm

$$\|S(t)\|_{q,p} := \sup_{\|f\|_{L^p(\mathcal{X})} \leq 1} \left(\|S(t) f\|_{L^q(\mathcal{X})} \right).$$

Of course, we have $\|S(t) f\|_{L^q(\mathcal{X})} \leq \|S(t)\|_{q,p} \|f\|_{L^p(\mathcal{X})}$, for all $t > 0$ and $f \in L^p(\mathcal{X})$, and

$$\|S(t)\|_{q,p} \leq C t^{-\beta_A \left(\frac{1}{p} - \frac{1}{q}\right)}, \quad \text{for all } t > 0.$$

Therefore, on account of (2.2.4) and (2.2.5), we also see for all $t > 0$,

$$\|S_\alpha(t)\|_{q,p} \leq C t^{-\beta_A \alpha \left(\frac{1}{p} - \frac{1}{q}\right)} \text{ and } \|t^{1-\alpha} P_\alpha(t)\|_{q,p} \leq C t^{-\beta_A \alpha \left(\frac{1}{p} - \frac{1}{q}\right)} \tag{2.2.12}$$

provided that $\beta_A (p^{-1} - q^{-1}) < 1$ and $\beta_A (p^{-1} - q^{-1}) < 2$, respectively, for any (fixed) $1 \leq p \leq q \leq \infty$.

The following continuity result will be used in Sect. 3.

Proposition 2.2.7 *Let* (2.2.3) *be satisfied by the semigroup S. Let $\theta \in (0, 1]$ and $p \in (1, \infty)$ such that $\theta p > \beta_A$. Then $D((-A_p)^\theta) \hookrightarrow L^\infty(\mathcal{X})$.*

Proof It suffices to show that the operator $(I - A_p)^{-\theta}$ maps $L^p(\mathcal{X})$ into $L^\infty(\mathcal{X})$ continuously. The assertion follows from the ultracontractivity estimate (2.2.3) and the well-known formula

$$(I - A_p)^{-\theta} = \frac{1}{\Gamma(\theta)} \int_0^\infty t^{\theta - 1} e^{-t} S(t)\, dt.$$

The proof is finished. □

2.3 Examples of Sectorial Operators

In this section, we give a sufficiently large number of examples of "diffusion" operators that satisfy the assumptions of the previous section (in particular, the statements of Propositions 2.2.1, 2.2.2). Many classical operators, that include uniformly (second-order) elliptic operators with sufficiently smooth coefficients defined on a smooth domain $\Omega \subset \mathbb{R}^N$ and subject to classical Dirichlet, Neumann or/and Robin boundary conditions, can be considered already by this analysis (see, for instance, [55]). However, the classical mathematical literature does not place much emphasis on the regularity of the domain and its immediate affect on the ultracontractivity estimates, although this kind of issues have been considered only recently (see, for instance, [9, 10, 63]; cf. also [34] and references therein). Most of our "diffusion" operators below take also into account a weaker regularity assumption on the domain, by allowing Ω to be non-smooth, as well as, we aim to present a number of recent examples that involve diffusion operators of "fractional" type; our goal is also to place more emphasis on the later kind. Nevertheless, we point out once again that the mapping assumptions we are going to employ out of Sect. 2.1 are more general, and as a result do not require any specific form of the diffusion operator. Such an abstraction allows us to represent a much larger family of fractional kinetic models that have not been explicitly studied in detail so far to the best of our knowledge.

But first, we need to introduce some general classes of Sobolev spaces. Let $\Omega \subset \mathbb{R}^N$ be an arbitrary bounded open set. Let

$$W^{1,2}(\Omega) := \left\{u \in L^2(\Omega) : \int_\Omega |\nabla u|^2 dx < \infty\right\}$$

be the first order Sobolev space endowed with the norm defined by

$$\|u\|_{W^{1,2}(\Omega)} = \left(\int_\Omega |u|^2 dx + \int_\Omega |\nabla u|^2 dx\right)^{\frac{1}{2}}.$$

We also let

$$W_0^{1,2}(\Omega) := \overline{\mathcal{D}(\Omega)}^{W^{1,2}(\Omega)},$$

where $\mathcal{D}(\Omega)$ denotes the space of all infinitely continuous differentiable functions with compact support in Ω. It is well-known that

$$W_0^{1,2}(\Omega) \overset{d}{\hookrightarrow} L^q(\Omega), \tag{2.3.1}$$

with

$$\begin{cases} 1 \le q \le \frac{2N}{N-2} & \text{if } N > 2, \\ 1 \le q < \infty & \text{if } N = 2, \\ 1 \le q \le \infty & \text{if } N = 1. \end{cases} \tag{2.3.2}$$

Remark 2.3.1 If Ω has the $W^{1,2}$-extension property, that is, for every $u \in W^{1,2}(\Omega)$, there exists $U \in W^{1,2}(\mathbb{R}^N)$ such that $U|_\Omega = u$ a.e. on Ω, then

$$W^{1,2}(\Omega) \overset{d}{\hookrightarrow} L^q(\Omega) \tag{2.3.3}$$

with q as in (2.3.2).

Next, let

$$W := \left\{u \in W^{1,2}(\Omega) \cap C(\overline{\Omega}) : \int_{\partial\Omega} |u|^2 d\sigma < \infty\right\},$$

where σ denotes the restriction to $\partial\Omega$ of the $(N-1)$-dimensional Hausdorff measure $\mathcal{H}^{N-1}$. The Maz'ya space $W_2^{1,2}(\Omega, \partial\Omega)$ is defined to be the completion of W with respect to the norm

$$\|u\|_{W_2^{1,2}(\Omega,\partial\Omega)} := \left(\int_\Omega |\nabla u|^2 dx + \int_{\partial\Omega} |u|^2 d\sigma\right)^{\frac{1}{2}}.$$

We have that

$$W_2^{1,2}(\Omega, \partial\Omega) \subset L^q(\Omega), \tag{2.3.4}$$

with

$$1 \le q \le \frac{2N}{N-1} \quad \text{if } N > 1 \text{ and } 1 \le q \le \infty \text{ if } N = 1. \tag{2.3.5}$$

Remark 2.3.2 Firstly, we notice that the inclusion (2.3.4) is continuous but is not always an injection. The latter property requires a regularity of the open set. For this reason, in (2.3.4) we did not use the notation $W_2^{1,2}(\Omega, \partial\Omega) \overset{d}{\hookrightarrow} L^q(\Omega)$. Secondly, if Ω has a Lipschitz continuous boundary, then $W_2^{1,2}(\Omega, \partial\Omega) = W^{1,2}(\Omega)$ and hence, (2.3.4) holds with q as in (2.3.2).

We refer to [17, 18, 49, 50] for a complete description of the Maz'ya type spaces. Finally for $0 < s < 1$, we introduce the fractional order Sobolev space

$$W^{s,2}(\Omega) := \left\{u \in L^2(\Omega) : \int_\Omega \int_\Omega \frac{|u(x) - u(y)|^2}{|x - y|^{N+2s}} dxdy < \infty\right\}$$

endowed with the norm

$$\|u\|_{W^{s,2}(\Omega)} = \left(\int_\Omega |u|^2 dx + \int_\Omega \int_\Omega \frac{|u(x) - u(y)|^2}{|x - y|^{N+2s}} dxdy\right)^{\frac{1}{2}}.$$

We let $W_0^{s,2}(\Omega) := \overline{\mathcal{D}(\Omega)}^{W^{s,2}(\Omega)}$, and

$$W_0^{s,2}(\overline{\Omega}) := \left\{u \in W^{s,2}(\mathbb{R}^N) : u = 0 \text{ on } \mathbb{R}^N \setminus \Omega\right\},$$

and

$$\widetilde{W}_0^{s,2}(\Omega) := \left\{u|_\Omega : u \in W_0^{s,2}(\overline{\Omega})\right\}.$$

Here also we have that

$$W_0^{s,2}(\Omega),\ \widetilde{W}_0^{s,2}(\Omega) \hookrightarrow L^q(\Omega), \tag{2.3.6}$$

with

$$\begin{cases} 1 \le q \le \frac{2N}{N-2s} & \text{if } N > 2s,\\ 1 \le q < \infty & \text{if } N = 2s\\ 1 \le q \le \infty & \text{if } N < 2s. \end{cases} \tag{2.3.7}$$

Remark 2.3.3 If Ω has the $W^{s,2}$-extension property, that is, for every $u \in W^{s,2}(\Omega)$, there exists $U \in W^{s,2}(\mathbb{R}^N)$ such that $U|_\Omega = u$ a.e. on Ω, then (2.3.6) holds with $W_0^{s,2}(\Omega)$ replaced with $W^{s,2}(\Omega)$.

We also observe the following facts.

Remark 2.3.4 Let $\Omega \subset \mathbb{R}^N$ be a bounded domain. Then we have the following situations.

(a) $D(\Omega) \subset W_0^{s,2}(\overline{\Omega})$ but is not always a dense subspace. If Ω has a continuous boundary, then $\mathcal{D}(\Omega)$ is dense in $W_0^{s,2}(\overline{\Omega})$. As a direct consequence, we have that $D(\Omega) \subset \widetilde{W}_0^{s,2}(\Omega)$ and is dense under the assumption that Ω has a continuous boundary.

(b) The spaces $W_0^{s,2}(\Omega)$ and $\widetilde{W}_0^{s,2}(\Omega)$ do not always coincide.

(c) Assume that Ω has a Lipschitz continuous boundary $\partial\Omega$. Then, by [38, Corollary 1.4.4.10] for every $0 < s < 1$,

$$\widetilde{W}_0^{s,2}(\Omega) = \left\{u \in W_0^{s,2}(\Omega) : \frac{u}{\delta^s} \in L^2(\Omega)\right\}, \tag{2.3.8}$$

where $\delta(x) := \text{dist}(x, \partial\Omega)$, $x \in \Omega$. By [38, Corollary 1.4.4.5] if $s \neq \frac{1}{2}$, then $W_0^{s,2}(\Omega) = \widetilde{W}_0^{s,2}(\Omega)$. But if $s = \frac{1}{2}$, then $\widetilde{W}_0^{s,2}(\Omega)$ is a proper subspace of $W_0^{\frac{1}{2},2}(\Omega)$. Notice also that $W_0^{s,2}(\Omega) = W^{s,2}(\Omega)$ for every $0 < s \le \frac{1}{2}$.

We refer to [2, 29, 38, 63] for a complete description and further properties of fractional order Sobolev spaces.

Our first example is the Laplace operator with various boundary conditions.

Example 2.3.5 (The Laplace Operator with Various Boundary Conditions) Let $\Omega \subset \mathbb{R}^N$ be an arbitrary bounded open set. All the considered bilinear forms are symmetric, closed, continuous and elliptic.

(a) **The Dirichlet boundary condition**. Let $\mathcal{E}_D$ with $D(\mathcal{E}_D) := W_0^{1,2}(\Omega)$ be given by

$$\mathcal{E}_D(u, v) := \int_\Omega \nabla u \cdot \nabla v dx. \tag{2.3.9}$$

Let Δ_D be the self-adjoint operator on $L^2(\Omega)$ associated with $\mathcal{E}_D$. Then Δ_D is a realization in $L^2(\Omega)$ of $-\Delta$ with the Dirichlet boundary condition. Using an integrating by parts argument, it is classical to show that

$$D(\Delta_D) = \left\{u \in W_0^{1,2}(\Omega),\ \Delta u \in L^2(\Omega)\right\},\ \Delta_D u = -\Delta u.$$

It follows from the coercivity of the form (2.3.9) and the embedding (2.3.1) with the value of q given in (2.3.2) that $\mathcal{E}_D$ satisfies the estimate (2.2.9) with $\mu = N$ if $N > 2$ and $\mu > 2$ an arbitrary real number if $N \leq 2$. Then Propositions 2.2.1 and 2.2.2 hold with $A = -\Delta_D$ and in that case $\beta_A = \frac{N}{2}$ if $N > 2$ and $\beta_A > 1$ arbitrary if $N \leq 2$. We also have that

$$\mathcal{L}^\infty(\Omega) = C_0(\Omega) := \{u \in C(\overline{\Omega}) :\ u = 0 \text{ on } \partial\Omega\},$$

if and only if Ω is regular in the sense of Wiener. If Ω is not Wiener regular, then we have that $C_0(\Omega) \subsetneq \mathcal{L}^\infty(\Omega) \subsetneq L^\infty(\Omega)$. We refer to [5, 6, 16] and their references for more details on this topic.

(b) **The Neumann boundary conditions**. Assume that Ω has the $W^{1,2}$-extension property in the sense of Remark 2.3.1. Let $\mathcal{E}_\mathcal{N}$ with $D(\mathcal{E}_\mathcal{N}) := W^{1,2}(\Omega)$ be defined by

$$\mathcal{E}_\mathcal{N}(u, v) := \int_\Omega \nabla u \cdot \nabla v dx. \tag{2.3.10}$$

Let $\Delta_\mathcal{N}$ be the self-adjoint operator on $L^2(\Omega)$ associated with $\mathcal{E}_\mathcal{N}$ in the sense of (2.2.7). Then $\Delta_\mathcal{N}$ is a realization in $L^2(\Omega)$ of $-\Delta$ with the Neumann boundary conditions. Using (2.3.3), we get that $\mathcal{E}_\mathcal{N} + (\cdot, \cdot)_{L^2(\Omega)}$ satisfies (2.2.9) with $\mu = N$ if $N > 2$ and $\mu > 2$ arbitrary if $N \leq 2$. Hence, Propositions 2.2.1 and 2.2.2 hold with $A = -\Delta_\mathcal{N} - I$ in which case $\beta_A = \frac{N}{2}$ if $N > 2$ and $\beta_A > 1$ arbitrary if $N \leq 2$. If Ω has a Lipschtiz continuous boundary, then an integration by parts argument gives that

$$\begin{cases} D(\Delta_{\mathcal{N}}) = \{u \in W^{1,2}(\Omega) : \Delta u \in L^2(\Omega) \text{ and } \partial_\nu u = 0 \text{ on } \partial\Omega\}, \\ \Delta_{\mathcal{N}} u = -\Delta u, \end{cases}$$

where $\partial_\nu u := \nabla u \cdot \nu$ denotes the normal derivative of u. In addition, in that case we have that $\mathcal{L}^\infty(\Omega) = C(\overline{\Omega})$ (see e.g. [61] and the references therein).

(c) **The Robin boundary conditions**. Let $\beta \in L^\infty(\partial\Omega)$ satisfy $\beta(x) \geq \beta_0 > 0$ for σ-a.e. $x \in \partial\Omega$ and let $\mathcal{E}_\beta$ be the bilinear and symmetric form defined by

$$\mathcal{E}_\beta(u, v) = \int_\Omega \nabla u \cdot \nabla v dx + \int_{\partial\Omega} \beta(x) uv d\sigma, \quad u, v \in W^{1,2}(\Omega) \cap C(\overline{\Omega}).$$

It is clear that $\mathcal{E}_\beta$ is not closed and by [9, 10] it is also not always closable, but it always has a closable part. Let $\mathcal{E}_R$ be the closure of the closable part of $\mathcal{E}_\beta$. It has been shown in [9, 10] that there exists a relatively closed set $\Gamma \subset \partial\Omega$ such that

$$D(\mathcal{E}_R) = \left\{u \in \widetilde{W}^{1,2}(\Omega) : \tilde{u} \in L^2(\Gamma)\right\}$$

and

$$\mathcal{E}_R(u, v) := \int_\Omega \nabla u \cdot \nabla v dx + \int_\Gamma \beta(x) \tilde{u}\tilde{v} d\sigma. \tag{2.3.11}$$

Here $\widetilde{W}^{1,2}(\Omega) = \overline{W^{1,2}(\Omega) \cap C(\overline{\Omega})}^{W^{1,2}(\Omega)}$ and $\tilde{u}$ denotes the relatively quasi-continuous representative (with respect to the capacity defined on subsets of $\overline{\Omega}$ with the regular Dirichlet space $\widetilde{W}^{1,2}(\Omega)$) of the function u. We also have that $D(\mathcal{E}_R) \hookrightarrow L^q(\Omega)$ with q as in (2.3.5) if Ω is arbitrary and as in Remark 2.3.2 if Ω has a Lipschitz continuous boundary. In addition $\mathcal{E}_R$ is closed, symmetric, continuous and coercive. Let Δ_R be the self-adjoint operator on $L^2(\Omega)$ associated with $\mathcal{E}_R$ in the sense of (2.2.7). Then Δ_R is a realization in $L^2(\Omega)$ of $-\Delta$ with the Robin boundary conditions. If Ω is an arbitrary bounded open set, then it follows from the coercivity of the form (2.3.11) and the embedding (2.3.4) that $\mathcal{E}_R$ satisfies (2.2.9) with $\mu = 2N$ if $N > 1$ and $\mu > 2$ arbitrary if $N = 1$. In that case, Propositions 2.2.1 and 2.2.2 hold with $A = -\Delta_R$ and $\beta_A = N$ if $N > 1$ and $\beta_A > 1$ arbitrary if $N = 1$. If Ω has a Lipschitz continuous boundary, then $D(\mathcal{E}_R) = W^{1,2}(\Omega)$ (see Remark 2.3.2) and

$$\mathcal{E}_R(u, v) = \int_\Omega \nabla u \cdot \nabla v dx + \int_{\partial\Omega} \beta(x) uv \, d\sigma.$$

Therefore, by (2.3.3), we have that $\mathcal{E}_R$ satisfies (2.2.9) with $\mu = N$ if $N > 2$ and $\mu > 2$ arbitrary if $N \leq 2$. Hence, Propositions 2.2.1 and 2.2.2 hold with $A = -\Delta_R$ and $\beta_A = \frac{N}{2}$ if $N > 2$ and $\beta_A > 1$ arbitrary if $N \leq 2$. In addition we have that

$$\begin{cases} D(\Delta_R) = \{u \in W^{1,2}(\Omega) : \ \Delta u \in L^2(\Omega) \ \text{and} \ \partial_\nu u + \beta u = 0 \ \text{on} \ \partial\Omega\}, \\ \Delta_R u = -\Delta u. \end{cases}$$

Furthermore (always under the assumption that Ω has a Lipschitz continuous boundary) it has been shown in [61] that $\mathcal{L}^\infty(\Omega) = C(\overline{\Omega})$. We refer to [9, 10, 17, 18, 26] for more details.

(d) **The Wentzell boundary conditions**. Assume that Ω has a Lipschitz continuous boundary. Let $\beta \in L^\infty(\partial\Omega)$ be as in part (c), $\delta \in \{0, 1\}$ and

$$\mathbb{W}^{1,\delta,2}(\overline{\Omega}) := \left\{ U = (u, u|_{\partial\Omega}) : \ u \in W^{1,2}(\Omega) \ \text{and} \ \delta u|_{\partial\Omega} \in W^{1,2}(\partial\Omega) \right\},$$

be endowed with the norm

$$\|u\|_{\mathbb{W}^{1,\delta,2}(\overline{\Omega})} = \begin{cases} \left(\|u\|^2_{W^{1,2}(\Omega)} + \|u\|^2_{W^{1,2}(\partial\Omega)} \right)^{\frac{1}{2}} & \text{if } \delta = 1 \\ \left(\|u\|^2_{W^{1,2}(\Omega)} + \|u\|^2_{W^{\frac{1}{2},2}(\partial\Omega)} \right)^{\frac{1}{2}} & \text{if } \delta = 0. \end{cases}$$

Then

$$\mathbb{W}^{1,0,2}(\overline{\Omega}) \hookrightarrow L^q(\Omega) \times L^q(\partial\Omega), \tag{2.3.12}$$

with

$$1 \leq q \leq \frac{2(N-1)}{N-2} \ \text{ if } \ N > 2 \ \text{ and } 1 \leq q < \infty \ \text{ if } \ N \leq 2, \tag{2.3.13}$$

and

$$\mathbb{W}^{1,1,2}(\overline{\Omega}) \hookrightarrow L^q(\Omega) \times L^q(\partial\Omega), \tag{2.3.14}$$

with

$$1 \leq q \leq \frac{2N}{N-2} \ \text{ if } \ N > 2 \ \text{ and } 1 \leq q < \infty \ \text{ if } \ N \leq 2. \tag{2.3.15}$$

Let $\mathcal{E}_{\delta,W}$ with $D(\mathcal{E}_{\delta,W}) := \mathbb{W}^{1,\delta,2}(\overline{\Omega})$ be given by

$$\mathcal{E}_{\delta,W}(U, V) := \int_\Omega \nabla u \cdot \nabla v dx + \delta \int_{\partial\Omega} \nabla_\Gamma u \cdot \nabla_\Gamma v d\sigma + \int_{\partial\Omega} \beta(x) u v d\sigma. \tag{2.3.16}$$

Let $\Delta_{\delta,W}$ be the self-adjoint operator in $L^2(\Omega) \times L^2(\partial\Omega)$ associated with $\mathcal{E}_{\delta,W}$ in the sense of (2.2.7). Then $\Delta_{\delta,W}$ is a realization in $L^2(\Omega) \times L^2(\partial\Omega)$ of $\Big(-\Delta, -\Delta_\Gamma\Big)$ with the generalized Wentzell boundary conditions. More precisely, we have that

$$D(\Delta_{\delta,W}) = \Big\{(u, u|_\Gamma) \in \mathbb{W}^{1,\delta,2}(\overline{\Omega}) : \Delta u \in L^2(\Omega) \text{ and } \\ -\delta\Delta_\Gamma(u|_{\partial\Omega}) + \partial_\nu u + \beta(u|_{\partial\Omega}) \in L^2(\partial\Omega)\Big\},$$

and

$$\Delta_{\delta,W}(u, u|_\Gamma) = \Big(-\Delta u, -\delta\Delta_\Gamma(u|_{\partial\Omega}) + \partial_\nu u + \beta(u|_{\partial\Omega})\Big).$$

We notice that for $1 \le q \le \infty$, the space $L^q(\Omega) \times L^q(\partial\Omega)$ endowed with the norm

$$\|(f, g)\|_{L^q(\Omega)\times L^q(\partial\Omega} = \begin{cases} \left(\|f\|^q_{L^q(\Omega)} + \|g\|^q_{L^q(\partial\Omega)}\right)^{1/q} & \text{if } 1 \le q < \infty, \\ \max\{\|f\|_{L^\infty(\Omega)}, \|g\|_{L^\infty(\Omega)} & \text{if } q = \infty, \end{cases}$$

can be identified with $L^q(\overline{\Omega}, \mathfrak{m})$ where the measure $\mathfrak{m}$ on $\overline{\Omega}$ is defined for a measurable set $A \subset \overline{\Omega}$ by $\mathfrak{m}(A) = |\Omega \cap A| + \sigma(\partial\Omega \cap A)$. It follows from the coercivity of the form (2.3.16) and the embeddings (2.3.12) and (2.3.14) with q given in (2.3.13) and (2.3.15) that $\mathcal{E}_{0,W}$ and $\mathcal{E}_{1,W}$ satisfy (2.2.9) with $\mu = 2(N-1)$ if $N > 2$, $\mu > 2$ arbitrary if $N \le 2$, and with $\mu = N$ if $N \ge 2$, $\mu > 2$ arbitrary if $N \le 2$, respectively. We have shown that Propositions 2.2.1 and 2.2.2 hold with $A = -\Delta_{0,W}$ and with $A = -\Delta_{1,W}$. In addition we have that in both cases $\mathcal{L}^\infty(\overline{\Omega}) = \{U = (u, u|_{\partial\Omega}) : u \in C(\overline{\Omega})\} \cong C(\overline{\Omega})$. For more details we refer to [61] and the references therein.

In all cases described by Example 2.3.5 one may replace the Laplace operator $-\Delta$ by a general second order elliptic operator L of the form

$$Lu = -\sum_{i,j=1}^{N} D_i\left(a_{ij}(x)D_j u\right), \tag{2.3.17}$$

where the coefficients $a_{ij} \in L^\infty(\Omega)$, $a_{ij}(x) = a_{ji}(x)$ for all $i, j = 1, 2, \dots, N$ and a.e. $x \in \Omega$, and

$$\sum_{i,j=1}^{N} a_{ij}(x)\xi_i\xi_j \ge \nu|\xi|^2, \ \forall\, \xi \in \mathbb{R}^N \text{ and for some constant } \nu > 0.$$

All the above results remain true when $-\Delta$ is replaced by L with the appropriate modifications.

Next we consider realizations of the fractional Laplace operator with various boundary conditions. For $0 < s < 1$,

$$u \in \mathcal{L}_s(\mathbb{R}^N) := \left\{u : \mathbb{R}^N \to \mathbb{R} \text{ measurable and } \int_{\mathbb{R}^N} \frac{|u(x)|}{(1+|x|)^{N+2s}} dx < \infty\right\},$$

and $\varepsilon > 0$ we let

$$(-\Delta)^s_\varepsilon u(x) := C_{N,s} \int_{\{y\in\mathbb{R}^N:\ |x-y|>\varepsilon\}} \frac{u(x)-u(y)}{|x-y|^{N+2s}} dy, \quad x \in \mathbb{R}^N,$$

and we define the **fractional Laplace operator** $(-\Delta)^s u$ as the following singular integral

$$(-\Delta)^s u(x) := C_{N,s}\text{P.V.} \int_{\mathbb{R}^N} \frac{u(x)-u(y)}{|x-y|^{N+2s}} dy = \lim_{\varepsilon\downarrow 0} (-\Delta)^s_\varepsilon u(x), \quad x \in \mathbb{R}^N, \tag{2.3.18}$$

provided that the limit exists for a.e. $x \in \mathbb{R}^N$, where the normalization constant $C_{N,s}$ is given by

$$C_{N,s} := \frac{s2^{2s}\Gamma\left(\frac{N}{2}+s\right)}{\pi^{\frac{N}{2}}\Gamma(1-s)}.$$

If, for a given function u, $(-\Delta)^s u \in L^2(\mathbb{R}^N)$, then we can let

$$(-\Delta)^s u := \frac{1}{\Gamma(-s)} \int_0^\infty \left(e^{t\Delta}u - u\right) \frac{dt}{t^{1+s}}$$

where $\Gamma(-s) = -\frac{\Gamma(1-s)}{s}$ and $(e^{t\Delta})_{t\geq 0}$ is the semigroup on $L^2(\mathbb{R}^N)$ generated by Δ. That is, we can define $(-\Delta)^s$ as the fractional s-power of the classical Laplace operator $-\Delta$. Furthermore, $(-\Delta)^s$ can be also defined as the pseudo-differential operator with symbol $|\xi|^{2s}$ by using Fourier transforms.
Throughout the following we shall write $(-\Delta)^s u \in L^2(\Omega)$ if the limit in (2.3.18) exists almost everywhere, and the function $x \mapsto (-\Delta)^s u(x)$ belongs to $L^2(\Omega)$.

Let $\Omega \subset \mathbb{R}^N$ be a bounded open set. For $u \in \mathcal{L}_s(\Omega)$ and $\varepsilon > 0$, we let

$$(-\Delta)^s_{\Omega,\varepsilon} u(x) = C_{N,s} \int_{\{y\in\Omega\ |x-y|>\varepsilon\}} \frac{u(x)-u(y)}{|x-y|^{N+2s}} dy, \quad x \in \Omega.$$

We set

$$(-\Delta)^s_\Omega u(x) = C_{N,s}\text{P.V.}\int_\Omega \frac{u(x)-u(y)}{|x-y|^{N+2s}}dy = \lim_{\varepsilon\downarrow 0}(-\Delta)^s_{\Omega,\varepsilon}u(x), \quad x\in\Omega, \tag{2.3.19}$$

provided that the limit exists for a.e. $x \in \Omega$, which is called the **regional fractional Laplace operator**. We notice that $\mathcal{L}_s(\mathbb{R}^N)$ and $\mathcal{L}_s(\Omega)$ are the right spaces on which $(-\Delta)^s_\varepsilon u(x)$ and $(-\Delta)^s_{\Omega,\varepsilon}u(x)$ exist for every $\varepsilon > 0$, respectively, and are continuous at the continuity points of u (see e.g. [19, 39, 40, 64] and their references).

Let us notice that even on the space $\mathcal{D}(\Omega)$ of test functions, the operator $(-\Delta)^s$ and $(-\Delta)^s_\Omega$ are different. More precisely, a simple calculation shows that for $u \in \mathcal{D}(\Omega)$ we have

$$(-\Delta)^s u(x) = (-\Delta)^s_\Omega u(x) + \kappa(x)u(x), \quad x\in\Omega,$$

where the function κ is given by

$$\kappa(x) := \int_{\mathbb{R}^N\setminus\Omega}\frac{1}{|x-y|^{N+2s}}\,dy, \quad x\in\Omega.$$

For more details we refer to the above references.

Example 2.3.6 (Fractional Laplace Operator with Various Exterior Conditions) Let $\Omega \subset \mathbb{R}^N$ be an arbitrary bounded open set. Here also all the bilinear forms that we will consider are symmetric, closed, continuous and elliptic.

(a) **The fractional Laplacian with Dirichlet exterior condition**. Before given our operator, let us first recall the following integration by parts formula. Let $u \in W_0^{s,2}(\overline{\Omega})$ be such that $(-\Delta)^s u \in L^2(\Omega)$. Then, for every $v \in W_0^{s,2}(\overline{\Omega})$ the identity

$$\int_\Omega v(-\Delta)^s u\,dx = \frac{C_{N,s}}{2}\int_{\mathbb{R}^N}\int_{\mathbb{R}^N}\frac{(u(x)-u(y))(v(x)-v(y))}{|x-y|^{N+2s}}dxdy \tag{2.3.20}$$

holds.

Next, we consider the following Dirichlet exterior value problem:

$$(-\Delta)^s u = f \quad \text{in } \Omega, \quad u = 0 \quad \text{in } \mathbb{R}^N\setminus\Omega. \tag{2.3.21}$$

Let $f \in L^2(\Omega)$. By a weak solution of (2.3.21) we mean a function $u \in W_0^{s,2}(\overline{\Omega})$ such that

$$\frac{C_{N,s}}{2}\int_{\mathbb{R}^N}\int_{\mathbb{R}^N}\frac{(u(x)-u(y))(v(x)-v(y))}{|x-y|^{N+2s}}dxdy = \int_\Omega fv\,dx$$

for all $v \in W_0^{s,2}(\overline{\Omega})$. Using the well-know Lax-Milgram theorem, it is straightforward to show that the Dirichlet problem (2.3.21) has a unique weak solution.

Let $u \in W_0^{s,2}(\overline{\Omega})$ be a weak solution of the Dirichlet problem (2.3.21). We have the following situation.

- We do not know if u is a strong solution of (2.3.21) in the sense that $(-\Delta)^s u$ (as defined in (2.3.19)) is well-defined almost everywhere and that $(-\Delta)^s u = f$ a.e. in Ω. Such a result always holds in the classical case of the Laplace operator.
- Here we just know that $u \in W_{\mathrm{loc}}^{2s,2}(\Omega)$ (see e.g. [15]). But this maximal inner regularity result is not enough to show that weak and strong solutions coincide. Let us notice that strong solutions are always weak solutions (this follows directly from the integration by parts formula (2.3.20)) as in the classical case $s = 1$. For this reason the operators we shall define are selfadjoint realizations of the fractional Laplace operator.

Let $\mathcal{E}_{s,D}$ with $D(\mathcal{E}_{s,D}) := \widetilde{W}_0^{s,2}(\Omega)$ be given by

$$\mathcal{E}_{s,D}(u, v) := \frac{C_{N,s}}{2} \int_{\mathbb{R}^N} \int_{\mathbb{R}^N} \frac{(u_D(x) - u_D(y))(v_D(x) - v_D(y))}{|x - y|^{N+2s}} dxdy,$$

where for a function $w \in L^2(\Omega)$ we have denoted

$$w_D := \begin{cases} w & \text{in } \Omega \\ 0 & \text{in } \mathbb{R}^N \setminus \Omega, \end{cases}$$

and we recall that

$$\widetilde{W}_0^{s,2}(\Omega) := \left\{ u|_{\Omega} : \ u \in W_0^{s,2}(\overline{\Omega}) \right\}.$$

Let $(-\Delta)_D^s$ be the selfadjoint operator on $L^2(\Omega)$ associated with $\mathcal{E}_{s,D}$ in the sense of (2.2.8). It has been shown in [23] that

$$\begin{cases} D((-\Delta)_D^s) = \Big\{ u \in \widetilde{W}_0^{s,2}(\Omega) : \ \exists\, f \in L^2(\Omega) \ \text{such that } u_D \text{ is a weak solution} \\ \qquad\qquad\qquad\qquad \text{of (2.3.21) with right hand side } f \Big\}, \\ (-\Delta)_D^s u = f. \end{cases}$$

Then $(-\Delta)_D^s$ is the realization in $L^2(\Omega)$ of $(-\Delta)^s$ with the Dirichlet exterior condition $u_D = 0$ on $\mathbb{R}^N \setminus \Omega$. It follows from (2.3.6) with q given in (2.3.7) that $\mathcal{E}_{s,D}$ satisfies (2.2.9) with $\mu = \frac{N}{s}$ if $N > 2s$ and $\mu > 2$ arbitrary if $N \leq 2s$. Hence, Propositions 2.2.1 and 2.2.2 hold with $A = -(-\Delta)_D^s$ in which case $\beta_A = \frac{N}{2s}$ if $N > 2s$ and $\beta_A > 1$ arbitrary if $N \leq 2s$. If Ω is of class $C^{1,1}$, then using the regularity results obtained in [54] one can show that $\mathcal{L}^\infty(\Omega) = C_0(\Omega)$. Let us mention that differently from the classical case, the fractional Laplacian also naturally admits boundary explosive solution, see e.g. [1, 31, 66].

(b) **The fractional Laplacian with nonlocal Neumann exterior condition**. Let $\Omega \subset \mathbb{R}^N$ be a bounded open set with Lipschitz continuous boundary. For $0 < s < 1$ we define the Hilbert space (see [30, Proposition 3.1] for the proof)

$$W_\Omega^{s,2} := \left\{u : \mathbb{R}^N \to \mathbb{R} \text{ measurable and } \|u\|_{W_\Omega^{s,2}} < \infty\right\},$$

where

$$\|u\|^2_{W_\Omega^{s,2}} := \|u\|^2_{L^2(\Omega)} + \int\int_{\mathbb{R}^{2N}\setminus(\mathbb{R}^N\setminus\Omega)^2} \frac{|u(x)-u(y)|^2}{|x-y|^{N+2s}} dxdy,$$

and

$$\mathbb{R}^{2N} \setminus (\mathbb{R}^N \setminus \Omega)^2 = (\Omega \times \Omega) \cup (\Omega \times (\mathbb{R}^N \setminus \Omega)) \cup ((\mathbb{R}^N \setminus \Omega \times \Omega).$$

For $u \in W_\Omega^{s,2}$ we define the **nonlocal fractional normal derivative** of u as

$$\mathcal{N}_s u(x) := C_{N,s} \int_\Omega \frac{u(x)-u(y)}{|x-y|^{N+2s}} dy, \quad x \in \mathbb{R}^N \setminus \Omega, \tag{2.3.22}$$

provided that the integral exists.
We have the following integration by parts formula. Let $u \in W_\Omega^{s,2}$ be such that $(-\Delta)^s u \in L^2(\Omega)$ and $\mathcal{N}_s u \in L^2(\mathbb{R}^N \setminus \Omega)$. Then for every $v \in W_\Omega^{s,2} \cap L^2(\mathbb{R}^N \setminus \Omega)$ the identity

$$\begin{aligned}\int_\Omega v(-\Delta)^s u\, dx =& \frac{C_{N,s}}{2} \int\int_{\mathbb{R}^{2N}\setminus(\mathbb{R}^N\setminus\Omega)^2} \frac{(u(x)-u(y))(v(x)-v(y))}{|x-y|^{N+2s}} dxdy \\ &- \int_{\mathbb{R}^N\setminus\Omega} v\mathcal{N}_s u\, dx \end{aligned}\tag{2.3.23}$$

holds. For a function $u \in L^2(\Omega)$, we define it extension u_N on $\mathbb{R}^N$ as follows

$$u_N(x) := \begin{cases} u(x) & \text{if } x \in \Omega, \\ \dfrac{1}{\rho(x)} \displaystyle\int_\Omega \frac{u(y)}{|x-y|^{N+2s}} dy & \text{if } x \in \mathbb{R}^N \setminus \overline{\Omega}. \end{cases}$$

where

$$\rho(x) := \int_\Omega \frac{1}{|x-y|^{N+2s}} dy, \quad x \in \mathbb{R}^N \setminus \overline{\Omega}. \tag{2.3.24}$$

Since we have assumed that Ω is bounded, using the Hölder inequality, we have that for every $u \in L^2(\Omega)$ and a.e. $x \in \mathbb{R}^N \setminus \Omega$,

$$|u_N(x)| \le \frac{\|u\|_{L^1(\Omega)}}{\rho(x)} \frac{1}{(\mathrm{dist}(x,\Omega))^{N+2s}}.$$

Thus, u_N is well-defined for every $u \in L^2(\Omega)$. It follows from the definition of u_N, that $\mathcal{N}_s u_N = 0$ in $\mathbb{R}^N \setminus \overline{\Omega}$.
Next, for $f \in L^2(\Omega)$ we consider the following Neumann exterior value problem:

$$(-\Delta)^s u = f \quad \text{in } \Omega, \quad \mathcal{N}_s u = 0 \quad \text{in } \mathbb{R}^N \setminus \Omega. \tag{2.3.25}$$

Let $f \in L^2(\Omega)$. By a weak solution of (2.3.25) we mean a function $u \in W^{s,2}_\Omega$ such that

$$\frac{C_{N,s}}{2} \int\int_{\mathbb{R}^{2N}\setminus(\mathbb{R}^N\setminus\Omega)^2} \frac{(u(x)-u(y))(v(x)-v(y))}{|x-y|^{N+2s}} dxdy = \int_\Omega fv\, dx \tag{2.3.26}$$

for all $v \in W^{s,2}_\Omega$. Using the Lax-Milgram theorem, it is easy to see that the Neumann problem (2.3.25) has a weak solution.
Let $u \in W^{s,2}_\Omega$ be a weak solution of the Neumann problem (2.3.25). Here also we do not know if u is a strong solution of (2.3.25) in the sense described above. We just know that the exterior condition is satisfied a.e. in $\mathbb{R}^N\setminus\Omega$. Indeed, taking $v \in \mathcal{D}(\mathbb{R}^N \setminus \overline{\Omega})$ as a test function in (2.3.26) and calculation we get that (notice that $v = 0$ in Ω)

$$\begin{aligned}
0 =& \frac{C_{N,s}}{2} \int\int_{\mathbb{R}^{2N}\setminus(\mathbb{R}^N\setminus\Omega)^2} \frac{(u(x)-u(y))(v(x)-v(y))}{|x-y|^{N+2s}} dxdy \\
=& C_{N,s} \int_\Omega \int_{\mathbb{R}^N\setminus\Omega} \frac{(u(x)-u(y))(v(x)v(y))}{|x-y|^{N+2s}} dxdy \\
=& C_{N,s} \int_{\mathbb{R}^N\setminus\Omega} v(y) \left(\int_\Omega \frac{u(x)-u(y)}{|x-y|^{N+2s}} dx \right) dy \\
=& \int_{\mathbb{R}^N\setminus\Omega} v(y) \mathcal{N}_s u(y)\, dy.
\end{aligned}$$

Since $v \in \mathcal{D}(\mathbb{R}^N \setminus \overline{\Omega})$ was arbitrary, we can deduce that $\mathcal{N}_s u = 0$ a.e. in $\mathbb{R}^N \setminus \Omega$. In addition, contrarily to the Dirichlet case, here we even do not know if weak solutions belong to $W^{2s,2}_{\mathrm{loc}}(\Omega)$. As in the Dirichlet case, here also strong solutions are always weak solutions. This also follows from the integration by parts formula (2.3.23).

Now let $\mathcal{E}_{s,\mathcal{N}}$ be the form on $L^2(\Omega)$ with domain

$$D(\mathcal{E}_{s,\mathcal{N}}) := \{u \in L^2(\Omega) : \; u_N \in W_\Omega^{s,2}\}$$

and given by

$$\mathcal{E}_{s,\mathcal{N}}(u,v) := \frac{C_{N,s}}{2} \int\int_{\mathbb{R}^{2N}\setminus(\mathbb{R}^N\setminus\Omega)^2} \frac{(u_N(x)-u_N(y))(v_N(x)-v_N(y))}{|x-y|^{N+2s}} dxdy.$$

Let us notice that it has been shown in [23] that

$$D(\mathcal{E}_{s,\mathcal{N}}) := \left\{u|_\Omega : \; u \in W_\Omega^{s,2}\right\}$$

and

$$\mathcal{E}_{s,\mathcal{N}}(u,u) := \inf\left\{\mathcal{E}(v,v) : \; v|_\Omega = u, \;\; v \in W_\Omega^{s,2}\right\},$$

where for $v, w \in W_\Omega^{s,2}$, we have set

$$\mathcal{E}(v,w) := \frac{C_{N,s}}{2} \int\int_{\mathbb{R}^{2N}\setminus(\mathbb{R}^N\setminus\Omega)^2} \frac{(v(x)-v(y))(w(x)-w(y))}{|x-y|^{N+2s}} dxdy.$$

It is easy to see that $D(\mathcal{E}_{s,\mathcal{N}}) \hookrightarrow W^{s,2}(\Omega)$. Hence, it follows from Remark 2.3.3 that $D(\mathcal{E}_{s,\mathcal{N}}) \hookrightarrow L^q(\Omega)$ with q as in (2.3.7). Let $(-\Delta)^s_{\mathcal{N}}$ be the selfadjoint operator on $L^2(\Omega)$ associated with $\mathcal{E}_{s,\mathcal{N}}$ in the sense of (2.2.8). It has been shown in [23] (see also [30, Section 3]) that

$$\begin{cases} D((-\Delta)^s_{\mathcal{N}}) = \Big\{u \in L^2(\Omega) : \; u_N \in W_\Omega^{s,2} : \; \exists\, f \in L^2(\Omega) \text{ such that } u_N \text{ is a} \\ \qquad\qquad \text{weak solution of (2.3.25) with right hand side } f\Big\}, \\ (-\Delta)^s_{\mathcal{N}} u = f. \end{cases}$$

Then $(-\Delta)^s_{\mathcal{N}}$ is the realization in $L^2(\Omega)$ of $(-\Delta)^s$ with the nonlocal fractional Neumann exterior condition $\mathcal{N}_s u_N = 0$ on $\mathbb{R}^N\setminus\overline{\Omega}$. Notice that the form $\mathcal{E}_{s,\mathcal{N}}$ is not coercive since the constant function $1 \in D(\mathcal{E}_{s,\mathcal{N}})$ and clearly $\mathcal{E}_{s,\mathcal{N}}(1,1) = 0$. Instead we have that $\mathcal{E}_{s,\mathcal{N}} + (\cdot,\cdot)_{L^2(\Omega)}$ is coercive and hence, satisfies (2.2.9) with $\mu = \frac{N}{s}$ if $N > 2s$ and $\mu > 2$ arbitrary if $N \le 2s$. This implies that Propositions 2.2.1 and 2.2.2 hold with $A = -(-\Delta)^s_{\mathcal{N}} - I$ in which case $\beta_A = \frac{N}{2s}$ if $N > 2s$ and $\beta_A > 1$ arbitrary if $N \le 2s$. Since we do not know if weak and strong solutions of the Neumann problem (2.3.25) coincide, we cannot deduce that $(-\Delta)^s_{\mathcal{N}}u = (-\Delta)^s u_N)|_\Omega$ for every $u \in D((-\Delta)^s_{\mathcal{N}})$. For more details on the operator $(-\Delta)^s_{\mathcal{N}}$ and the nonlocal fractional normal derivative given in (2.3.22) we refer the interested reader to [23, 30] and their references.

(c) **The fractional Laplacian with nonlocal Robin exterior condition**. Let $\Omega \subset \mathbb{R}^N$ be a bounded open set with Lipschitz continuous boundary and let $0 < s < 1$. Let $\beta \in L^1(\mathbb{R}^N \setminus \Omega)$ be a non-negative function. We define the fractional order Sobolev type space

$$W^{s,2}_{\beta,\Omega} := \left\{u \in W^{s,2}_\Omega : \int_{\mathbb{R}^N\setminus\Omega} \beta|u|^2\,dx < \infty\right\}$$

and we endow it with the norm given by

$$\|u\|_{W^{s,2}_{\beta,\Omega}} := \left(\|u\|^2_{W^{s,2}_\Omega} + \int_{\mathbb{R}^N\setminus\Omega} \beta|u|^2\,dx\right)^{\frac{1}{2}}.$$

It has been shown in [30, Proposition 3.1] that $W^{s,2}_{\beta,\Omega}$ is a Hilbert space.
Next we consider the following Robin exterior value problem:

$$(-\Delta)^s u = f \quad \text{in } \Omega, \quad \mathcal{N}_s u + \beta u = 0 \quad \text{in } \mathbb{R}^N \setminus \Omega. \tag{2.3.27}$$

Let $f \in L^2(\Omega)$. By a weak solution of (2.3.27) we mean a function $u \in W^{s,2}_{\beta,\Omega}$ such that

$$\begin{aligned}\frac{C_{N,s}}{2}\int\int_{\mathbb{R}^{2N}\setminus(\mathbb{R}^N\setminus\Omega)^2} &\frac{(u(x)-u(y))(v(x)-v(y))}{|x-y|^{N+2s}}dxdy + \int_{\mathbb{R}^N\setminus\Omega}\beta uv\,dx\\ &= \int_\Omega fv\,dx \end{aligned}\tag{2.3.28}$$

for all $v \in W^{s,2}_{\beta,\Omega}$. Here also using the Lax-Milgram theorem, it is straightforward to see that the Robin problem (2.3.27) has a unique weak solution.
Let $u \in W^{s,2}_{\beta,\Omega}$ be a weak solution of the Robin problem (2.3.27). We do not know if u is a strong solution of (2.3.27) in the sense described above, and if u belongs to $W^{2s,2}_{\text{loc}}(\Omega)$. As in the Neumann case we just know that the exterior condition in (2.3.27) is satisfied a.e. in $\mathbb{R}^N \setminus \Omega$. Indeed, taking $v \in \mathcal{D}(\mathbb{R}^N \setminus \overline{\Omega})$ as a test function in (2.3.28) and calculation we get that (notice that $v = 0$ in Ω)

$$\begin{aligned}0 =&\frac{C_{N,s}}{2}\int\int_{\mathbb{R}^{2N}\setminus(\mathbb{R}^N\setminus\Omega)^2}\frac{(u(x)-u(y))(v(x)-v(y))}{|x-y|^{N+2s}}dxdy + \int_{\mathbb{R}^N\setminus\Omega}\beta uv\,dy\\ =&C_{N,s}\int_\Omega\int_{\mathbb{R}^N\setminus\Omega}\frac{(u(x)-u(y))(v(x)v(y))}{|x-y|^{N+2s}}dxdy + \int_{\mathbb{R}^N\setminus\Omega}\beta uv\,dy\end{aligned}$$

$$=C_{N,s}\int_{\mathbb{R}^N\setminus\Omega} v(y)\left(\int_\Omega \frac{u(x)-u(y)}{|x-y|^{N+2s}}dx\right)dy+\int_{\mathbb{R}^N\setminus\Omega}\beta uv\,dy$$

$$=\int_{\mathbb{R}^N\setminus\Omega} v(y)\Big(\mathcal{N}_s u(y)+\beta(y)u(y)\Big)\,dy.$$

Since $v\in\mathcal{D}(\mathbb{R}^N\setminus\overline{\Omega})$ was arbitrary, we can deduce that $\mathcal{N}_s u+\beta u=0$ a.e. in $\mathbb{R}^N\setminus\Omega$.

Next, for a function $u\in L^2(\Omega)$, we define its extension u_R on $\mathbb{R}^N$ as follows:

$$u_R(x):=\begin{cases} u(x) & \text{if } x\in\Omega,\\ \dfrac{C_{n,s}}{C_{n,s}\rho(x)+\beta(x)}\displaystyle\int_\Omega \frac{u(y)}{|x-y|^{n+2s}}dy & \text{if } x\in\mathbb{R}^N\setminus\Omega,\end{cases}$$

where we recall that $\rho(x)$ is given by (2.3.24). As in the Neumann case, we have that u_R is well-defined for every $u\in L^2(\Omega)$. It has been shown in [23] that for every $u\in W^{s,2}_\Omega$, we have the equality.

$$\mathcal{N}^s u_R(x)+\beta(x)u_R(x)=0,\quad x\in\mathbb{R}^N\setminus\overline{\Omega}. \tag{2.3.29}$$

We call (2.3.29) the nonlocal Robin exterior condition.

Now we introduce the realization in $L^2(\Omega)$ of $(-\Delta)^s$ with the nonlocal Robin exterior condition. Let

$$D(\mathcal{E}_{s,R}):=\left\{u\in L^2(\Omega):\ u_R\in W^{s,2}_\Omega\ \text{ and }\ \int_{\mathbb{R}^N\setminus\Omega}\beta(x)u_R^2(x)\,dx<\infty\right\}$$

and $\mathcal{E}_{s,R}: D(\mathcal{E}_{s,R})\times D(\mathcal{E}_{s,R})\to\mathbb{R}$ given by

$$\mathcal{E}_{s,R}(u,v)=\frac{C_{N,s}}{2}\int\int_{\mathbb{R}^{2N}\setminus(\mathbb{R}^N\setminus\Omega)^2}\frac{(u_R(x)-u_R(y))(v_R(x)-v_R(y))}{|x-y|^{N+2s}}dxdy$$
$$+\int_{\mathbb{R}^N\setminus\Omega}\beta(x)u_R(x)v_R(x)\,dx.$$

Then $\mathcal{E}_{s,R}$ is a closed, symmetric and densely defined bilinear form on $L^2(\Omega)$. The selfadjoint operator $(-\Delta)^s_R$ associated with $\mathcal{E}_{s,R}$ (in the sense of (2.2.8)) is given by

$$\begin{cases} D((-\Delta)^s_R)=\Big\{u\in L^2(\Omega):\ u_R\in W^{s,2}_{\beta,\Omega}:\ \exists\, f\in L^2(\Omega)\ \text{such that } u_R \text{ is a}\\ \qquad\qquad\qquad \text{weak solution of (2.3.27) with right hand side } f\Big\},\\ (-\Delta)^s_R u=f.\end{cases}$$

It is easy to see that $D((-\Delta)_R^s) \hookrightarrow W^{s,2}(\Omega)$. Hence, it follows from Remark 2.3.3 that $D((-\Delta)_R^s) \hookrightarrow L^q(\Omega)$ with q as in (2.3.7). We notice that if $\beta \neq 0$ a.e. in $\mathbb{R}^N \setminus \Omega$, then the form $\mathcal{E}_{s,R}$ is in addition coercive. We mention that as in the Neumann case, since we do not know if weak and strong solutions of the Robin problem (2.3.27) coincide, we cannot conclude that $(-\Delta)_R^s u = (-\Delta)^s u_R)|_\Omega$ for every $u \in D((-\Delta)_R^s)$. For more details we refer to [3, 23].

(d) **The regional fractional Laplacian with Dirichlet boundary condition**. Let $\mathcal{E}_{\Omega,s,D}$ with $D(\mathcal{E}_{\Omega,s,D}) := W_0^{s,2}(\Omega)$ be defined by

$$\mathcal{E}_{\Omega,s,D}(u,v) := \frac{C_{N,s}}{2}\int_\Omega\int_\Omega \frac{(u(x)-u(y))(v(x)-v(y))}{|x-y|^{N+2s}}dxdy. \tag{2.3.30}$$

Let $(-\Delta)_{\Omega,D}^s$ be the self-adjoint operator on $L^2(\Omega)$ associated with $\mathcal{E}_{\Omega,s,D}$. Here also an integration by parts argument gives that

$$D((-\Delta)_{\Omega,D}^s) = \Big\{u \in W_0^{s,2}(\Omega) : (-\Delta)_\Omega^s u \in L^2(\Omega)\Big\},\ (-\Delta)_{\Omega,D}^s u = (-\Delta)_\Omega^s u.$$

The operator $(-\Delta)_{\Omega,D}^s$ is the realization in $L^2(\Omega)$ of $(-\Delta)_\Omega^s$ with the Dirichlet boundary condition $u = 0$ on $\partial\Omega$. It follows from (2.3.6) with q given in (2.3.7) that $\mathcal{E}_{\Omega,s,D}$ satisfies (2.2.9) with $\mu = \frac{N}{s}$ if $N > 2s$ and $\mu > 2$ arbitrary if $N \leq 2s$. Hence, Propositions 2.2.1 and 2.2.2 hold with $A = -(-\Delta)_{\Omega,D}^s$ in which case $\beta_A = \frac{N}{2s}$ if $N > 2s$ and $\beta_A > 1$ arbitrary if $N \leq 2s$. If Ω is of class $C^{1,1}$ and $\frac{1}{2} < s < 1$, then it follows from the results obtained in [39] that $\mathcal{L}^\infty(\Omega) = C_0(\Omega)$.

(e) **The regional fractional Laplacian with Neumann boundary conditions.** Assume that Ω has a Lipschitz continuous boundary. Let $\frac{1}{2} < s < 1$ and $\mathcal{E}_{\Omega,s,\mathcal{N}}$ with $D(\mathcal{E}_{\Omega,s,\mathcal{N}}) := W^{s,2}(\Omega)$ be given by

$$\mathcal{E}_{\Omega,s,\mathcal{N}}(u,v) := \frac{C_{N,s}}{2}\int_\Omega\int_\Omega \frac{(u(x)-u(y))(v(x)-v(y))}{|x-y|^{N+2s}}dxdy. \tag{2.3.31}$$

Let $(-\Delta)_{\Omega,\mathcal{N}}^s$ be the self-adjoint operator on $L^2(\Omega)$ associated with $\mathcal{E}_{\Omega,s,\mathcal{N}}$. It has been shown in [35, 62] by using the Green formula (B.0.7) that

$$\begin{cases} D((-\Delta)_{\Omega,\mathcal{N}}^s) = \{u \in W^{s,2}(\Omega) : (-\Delta)_\Omega^s u \in L^2(\Omega) \text{ and } \mathcal{N}^{2-2s}u = 0 \text{ on } \partial\Omega\}, \\ (-\Delta)_{\Omega,\mathcal{N}}^s u = (-\Delta)_\Omega^s u, \end{cases}$$

where $\mathcal{N}^{2-2s}u$ denotes the fractional normal derivative of u in the sense of Definition B.0.7 (see Appendix B below). The operator $(-\Delta)_{\Omega,\mathcal{N}}^s$ is the realization in $L^2(\Omega)$ of $(-\Delta)_\Omega^s$ with the fractional Neumann type boundary conditions. Using Remark 2.3.3 we get that $\mathcal{E}_{\Omega,s,\mathcal{N}} + (\cdot,\cdot)_{L^2(\Omega)}$ satisfies (2.2.9) with $\mu = \frac{N}{s}$ if $N > 2s$ and $\mu > 2$ arbitrary if $N \leq 2s$. This implies that Propositions 2.2.1 and 2.2.2 hold with $A = -(-\Delta)^s_{\Omega,\mathcal{N}} - I$ in which case $\beta_A = \frac{N}{2s}$ if $N > 2s$ and $\beta_A > 1$ arbitrary if $N \leq 2s$. We notice that the assumption that $\frac{1}{2} < s < 1$ is not a restriction. In fact, if $0 < s \leq \frac{1}{2}$, since $W_0^{s,2}(\Omega) = W^{s,2}(\Omega)$ (by Remark 2.3.4), we have that the forms $\mathcal{E}_{\Omega,s,D}$ and $\mathcal{E}_{\Omega,s,\mathcal{N}}$ given in (2.3.30) and (2.3.31), respectively, coincide. Thus, in this case Dirichlet and Neumann boundary conditions are the same.

(f) **The regional fractional Laplacian with Robin boundary conditions.** Assume that Ω has a Lipschitz continuous boundary and let $\beta \in L^\infty(\partial\Omega)$ satisfy $\beta(x) \geq \beta_0 > 0$. Let $\frac{1}{2} < s < 1$ and $\mathcal{E}_{\Omega,s,R}$ with $D(\mathcal{E}_{\Omega,s,R}) := W^{s,2}(\Omega)$ be given by

$$\mathcal{E}_{\Omega,s,R}(u,v) := \frac{C_{N,s}}{2}\int_\Omega\int_\Omega \frac{(u(x)-u(y))(v(x)-v(y))}{|x-y|^{N+2s}}dxdy + \int_{\partial\Omega}\beta(x)uv\,d\sigma.$$

Let $(-\Delta)^s_{\Omega,R}$ be the self-adjoint operator on $L^2(\Omega)$ associated with $\mathcal{E}_{\Omega,s,R}$. Using again the Green formula (B.0.7), it follows from [35] that

$$\begin{cases} D((-\Delta)^s_{\Omega,R}) = \{u \in W^{s,2}(\Omega) : \ (-\Delta)^s_\Omega u \in L^2(\Omega) \\ \qquad\qquad\qquad \text{and } \ C_s\mathcal{N}^{2-2s}u + \beta u = 0 \text{ on } \ \partial\Omega\}, \\ (-\Delta)^s_{\Omega,R}u = (-\Delta)^s_\Omega u, \end{cases}$$

where C_s is a normalization constant depending only on s (see Appendix B below). The operator $(-\Delta)^s_{\Omega,R}$ is the realization in $L^2(\Omega)$ of $(-\Delta)^s_\Omega$ with the fractional Robin type boundary conditions. Here also $\mathcal{E}_{\Omega,s,R}$ satisfies (2.2.9) with $\mu = \frac{N}{s}$ if $N > 2s$ and $\mu > 2$ arbitrary if $N \leq 2s$ and this implies that Propositions 2.2.1 and 2.2.2 hold with $A = -(-\Delta)^s_{\Omega,R}$ in which case $\beta_A = \frac{N}{2s}$ if $N > 2s$ and $\beta_A > 1$ arbitrary if $N \leq 2s$. As above, here also, the assumption $\frac{1}{2} < s < 1$ is not a restriction.

(g) **The regional fractional Laplacian with general Wentzell boundary conditions**. Let β, s and Ω be as in part (e). Let $\delta \in \{0, 1\}$ and

$$\mathbb{W}^{s,\delta,2}(\overline{\Omega}) := \Big\{U = (u, u|_{\partial\Omega}) : \ u \in W^{s,2}(\Omega) \text{ and } \delta u|_{\partial\Omega} \in W^{s,2}(\partial\Omega)\Big\},$$

be endowed with the norm

$$\|u\|_{\mathbb{W}^{s,\delta,2}(\overline{\Omega})} = \begin{cases} \left(\|u\|^2_{W^{s,2}(\Omega)} + \|u\|^2_{W^{s,2}(\partial\Omega)}\right)^{\frac{1}{2}} & \text{if } \delta = 1 \\ \left(\|u\|^2_{W^{s,2}(\Omega)} + \|u\|^2_{W^{s-\frac{1}{2},2}(\partial\Omega)}\right)^{\frac{1}{2}} & \text{if } \delta = 0. \end{cases}$$

Then

$$\mathbb{W}^{s,0,2}(\overline{\Omega}) \hookrightarrow L^q(\Omega) \times L^q(\partial\Omega), \tag{2.3.32}$$

with

$$1 \le q \le \frac{2(N-1)}{N-2s} \text{ if } N > 2s \text{ and } 1 \le q < \infty \text{ if } N \le 2s, \tag{2.3.33}$$

and

$$\mathbb{W}^{s,1,2}(\overline{\Omega}) \hookrightarrow L^q(\Omega) \times L^q(\partial\Omega), \tag{2.3.34}$$

with

$$1 \le q \le \frac{2N}{N-2s} \text{ if } N > 2s \text{ and } 1 \le q < \infty \text{ if } N \le 2s. \tag{2.3.35}$$

Let $\mathcal{E}_{s,\delta,W}$ with $D(\mathcal{E}_{\delta,W}) := \mathbb{W}^{s,\delta,2}(\overline{\Omega})$ be given by

$$\begin{aligned}\mathcal{E}_{s,\delta,W}(U,V) :=& \frac{C_{N,s}}{2}\int_\Omega\int_\Omega \frac{(u(x)-u(y))(v(x)-v(y))}{|x-y|^{N+2s}}dxdy + \int_{\partial\Omega}\beta(x)uvd\sigma \\ &+ \delta\frac{C_{N-1,s}}{2}\int_{\partial\Omega}\int_{\partial\Omega}\frac{(u(x)-u(y))(v(x)-v(y))}{|x-y|^{N-1+2s}}d\sigma_x d\sigma_y.\end{aligned}$$

Let $(-\Delta)^s_{\delta,W}$ be the self-adjoint operator on $L^2(\Omega) \times L^2(\partial\Omega)$ associated with $\mathcal{E}_{s,\delta,W}$ in the sense of (2.2.7). Then $(-\Delta)^s_{\delta,W}$ is a realization in $L^2(\Omega) \times L^2(\partial\Omega)$ of $\left((-\Delta)^s_\Omega, \delta(-\Delta)^s_\Gamma\right)$ with the generalized Wentzell boundary conditions. Here by $(-\Delta)^s_\Gamma$ we mean the operator defined formally for

$$v \in \mathcal{L}^s(\partial\Omega) := \left\{v : \partial\Omega \to \mathbb{R} \text{ measurable and } \int_{\partial\Omega}\frac{|v(x)|}{(1+|x|)^{N-1+2s}}\, d\sigma_x < \infty\right\},$$

by

$$(-\Delta)^s_\Gamma v := C_{N-1,s}\text{P.V.}\int_{\partial\Omega}\frac{v(x)-v(y)}{|x-y|^{N-1+2s}}\, d\sigma_y, \quad x \in \partial\Omega.$$

More precisely, we have that

$$\begin{aligned}D((-\Delta)^s_{\delta,W}) = \Big\{&U = (u, u|_\Gamma) \in \mathbb{W}^{s,\delta,2}(\overline{\Omega}) : \; (-\Delta)^s_\Omega u \in L^2(\Omega) \\ &\text{and } \delta(-\Delta)^s_\Gamma(u|_{\partial\Omega}) + C_s\mathcal{N}^{2-2s}u + \beta(u|_{\partial\Omega}) \in L^2(\partial\Omega)\Big\},\end{aligned}$$

and

$$(-\Delta)^s_{\delta,W}U = \Big((-\Delta)^s_\Omega u, \delta(-\Delta)^s_\Gamma(u|_{\partial\Omega}) + C_s N^{2-2s}u + \beta(u|_{\partial\Omega})\Big).$$

Using (2.3.32) with q given in (2.3.33) and (2.3.34) with q given in (2.3.35), we get that $\mathcal{E}_{s,0,W}$ and $\mathcal{E}_{s,1,W}$ satisfy (2.2.9) with $\mu = \frac{2(N-1)}{2s-1}$ If $N > 2s$, $\mu > 2$ arbitrary if $N \le 2s$, and $\mu = \frac{N}{s}$ if $N > 2s$, $\mu > 2$ arbitrary if $N \le 2s$, respectively. We have shown that Propositions 2.2.1 and 2.2.2 hold with $A = -(-\Delta)^s_{0,W}$ in which case $\beta_A = \frac{N-1}{2s-1}$ if $N > 2s$, $\beta_A > 1$ arbitrary if $N \le 2s$, and with $A = -(-\Delta)^s_{1,W}$ in which case $\beta_A = \frac{N}{2s}$ if $N > 2s$, $\beta_A > 1$ arbitrary if $N \le 2s$. We refer to [33] for more details on this topic.

For the cases (b), (c), (e), (f) and (g) in Example 2.3.6, there is no regularity result available in the literature that can be used to characterize the corresponding space $\mathcal{L}^\infty(\Omega)$ or $\mathcal{L}^\infty(\overline{\Omega})$.

We mention that in Example 2.3.6 parts (a) and (d), one may replace the kernel $|x-y|^{-N-2s}$ by a general symmetric kernel $K : \mathbb{R}^N \times \mathbb{R}^N \to [0,\infty)$ satisfying

$$C_1 \le K(x,y)|x-y|^{N+2s} \le C_2$$

for a.e. $x, y \in \mathbb{R}^N$ and for some constants $0 < C_1 \le C_2$. In that case our corresponding operators (2.3.18) and (2.3.19) are given by

$$A^s u(x) = C_{N,s}\text{P.V.}\int_{\mathbb{R}^N} K(x,y)(u(x)-u(y))dy$$

and

$$A^s_\Omega u(x) = C_{N,s}\text{P.V.}\int_\Omega K(x,y)(u(x)-u(y))dy,$$

respectively.

Next, we consider some Dirichlet-to-Neumann type operators.

Example 2.3.7 Throughout this example we assume that $\Omega \subset \mathbb{R}^N$ is a bounded open set with a Lipschitz continuous boundary $\partial\Omega$.

(a) **The classical Dirichlet-to-Neumann operator**. Recall that the operator Δ_D defined in Example 2.3.5(a) has a compact resolvent and its eigenvalues form a sequence of real numbers $0 < \lambda_1^D \le \lambda_2^D \le \cdots \le \lambda_n^D \cdots$ satisfying $\lim_{n\to\infty}\lambda_n^D = \infty$. We denote its spectrum by $\sigma(\Delta_D)$. Let $\lambda \in \mathbb{R}\backslash\sigma(\Delta_D)$, $g \in L^2(\partial\Omega)$ and let $u \in W^{1,2}(\Omega)$ be the weak solution of the Dirichlet problem

$$-\Delta u = \lambda u \text{ in } \Omega, \quad u|_{\partial\Omega} = g. \tag{2.3.36}$$

The classical Dirichlet to Neumann map is the operator $\mathbb{D}_{1,\lambda}$ on $L^2(\partial\Omega)$ with domain

$$D(\mathbb{D}_{1,\lambda}) = \Big\{ g \in L^2(\partial\Omega) : \exists\, u \in W^{1,2}(\Omega) \text{ solution of (2.3.36)} \\ \text{and } \partial_\nu u \text{ exists in } L^2(\partial\Omega) \Big\},$$

and given by

$$\mathbb{D}_{1,\lambda} g = \partial_\nu u.$$

It is well known that one has the following orthogonal decomposition

$$W^{1,2}(\Omega) = W_0^{1,2}(\Omega) \oplus \mathcal{H}^{1,\lambda}(\Omega),$$

where

$$\mathcal{H}^{1,\lambda}(\Omega) = \Big\{ u \in W^{1,2}(\Omega), \quad -\Delta u = \lambda u \Big\},$$

and by $-\Delta u = \lambda u$ we mean that

$$\int_\Omega \nabla u \cdot \nabla u dx = \lambda \int_\Omega u v dx, \ \forall\, v \in W_0^{1,2}(\Omega).$$

Let

$$W^{\frac{1}{2},2}(\partial\Omega) := \Big\{ u|_{\partial\Omega} : u \in W^{1,2}(\Omega) \Big\}$$

be the trace space. Since $\lambda \in \mathbb{R}\backslash\sigma(\Delta_D)$, we have that the trace operator restricted to $\mathcal{H}^{1,\lambda}(\Omega)$, that is, the mapping $u \in \mathcal{H}^{1,\lambda}(\Omega) \mapsto u|_{\partial\Omega} \in W^{\frac{1}{2},2}(\partial\Omega)$, is linear and bijective. Letting

$$\|u|_{\partial\Omega}\|_{W^{\frac{1}{2},2}(\partial\Omega)} = \|u\|_{\mathcal{H}^{1,\lambda}(\Omega)},$$

then $W^{\frac{1}{2},2}(\partial\Omega)$ becomes a Hilbert space. By the closed graph theorem, different choice of $\lambda \in \mathbb{R}\backslash\sigma(\Delta_D)$ leads to an equivalent norm on $W^{\frac{1}{2},2}(\partial\Omega)$. Moreover, we have that $W^{\frac{1}{2},2}(\partial\Omega) \overset{c}{\hookrightarrow} L^2(\partial\Omega)$ and $W^{\frac{1}{2},2}(\partial\Omega) \overset{d}{\hookrightarrow} L^2(\partial\Omega)$. In addition we have the continuous embedding

$$W^{\frac{1}{2},2}(\partial\Omega) \hookrightarrow L^q(\partial\Omega) \tag{2.3.37}$$

with

$$1 \le q \le \frac{2(N-1)}{N-2} \text{ if } N > 2 \text{ and } 1 \le q < \infty \text{ if } N \le 2. \tag{2.3.38}$$

It has been shown in [7] that $\mathbb{D}_{1,\lambda}$ is the self-adjoint operator on $L^2(\partial\Omega)$ associated with the bilinear, symmetric and continuous form $\mathcal{E}_{1,\lambda}$ with domain $W^{\frac{1}{2},2}(\partial\Omega)$ given by

$$\mathcal{E}_{1,\lambda}(\varphi, \psi) = \int_\Omega \nabla u \cdot \nabla v dx - \lambda \int_\Omega uv dx,$$

where $\varphi, \psi \in W^{\frac{1}{2},2}(\partial\Omega)$ and $u, v \in \mathcal{H}^{1,\lambda}(\Omega)$ are such that $u|_{\partial\Omega} = \varphi$ and $v|_{\partial\Omega} = \psi$. The operator $-\mathbb{D}_{1,\lambda}$ generates a strongly continuous and analytic semigroup on $L^2(\partial\Omega)$ which is also submarkovian if $\lambda \le 0$. If $\lambda < 0$ we also have that $\mathcal{E}_{1,\lambda}$ is coercive. In that case (by using (2.3.37) and (2.3.38)) we get that $\mathcal{E}_{1,\lambda}$ satisfies (2.2.9) with $\mathcal{X} = \partial\Omega$ and the constant $\mu = 2(N-1)$ if $N > 2$ and $\mu > 2$ arbitrary if $N \le 2$. Hence, Propositions 2.2.1 and 2.2.2 hold with $A = -\mathbb{D}_{1,\lambda}$ in which case $\beta_A = N-1$ if $N > 2$ and $\beta_A > 1$ arbitrary if $N \le 2$. In addition we have that $\mathcal{L}^\infty(\partial\Omega) = C(\partial\Omega)$. We refer to [7, 8, 12] and their references for more details on this topic.

(b) **The fractional Dirichlet-to-Neumann operator**. Let $\frac{1}{2} < s < 1$. We notice that the operator $(-\Delta)^s_{\Omega,D}$ defined in Example 2.3.6(b) has a compact resolvent and its eigenvalues form a sequence of real numbers $0 < \lambda_1^{s,D} \le \lambda_2^{s,D} \le \cdots \le \lambda_n^{s,D} \cdots$ satisfying $\lim_{n\to\infty} \lambda_n^{s,D} = \infty$. We denote its spectrum by $\sigma((-\Delta)^s_{\Omega,D})$. Let $\lambda \in \mathbb{R} \setminus \sigma((-\Delta)^s_{\Omega,D})$ be a real number, $g \in L^2(\partial\Omega)$ and let $u \in W^{s,2}(\Omega)$ be the weak solution of the following Dirichlet problem

$$(-\Delta)^s_\Omega u = \lambda u \text{ in } \Omega, \quad u|_{\partial\Omega} = g. \tag{2.3.39}$$

The fractional Dirichlet-to-Neumann map is the operator $\mathbb{D}_{s,\lambda}$ on $L^2(\partial\Omega)$ with domain

$$D(\mathbb{D}_{s,\lambda}) = \Big\{ g \in L^2(\partial\Omega) : \exists\, u \in W^{s,2}(\Omega) \text{ solution of (2.3.39)} \\ \text{and } \mathcal{N}^{2-2s}u \text{ exists in } L^2(\partial\Omega) \Big\},$$

and given by

$$\mathbb{D}_{s,\lambda} g = C_s \mathcal{N}^{2-2s} u,$$

where C_s is a normalized constant depending only on s (see Appendix B). One has the following orthogonal decomposition

$$W^{s,2}(\Omega) = W_0^{s,2}(\Omega) \oplus \mathcal{H}^{s,\lambda}(\Omega),$$

where

$$\mathcal{H}^{s,\lambda}(\Omega) = \Big\{u \in W^{s,2}(\Omega), \quad (-\Delta)^s_\Omega u = \lambda u\Big\},$$

and by $(-\Delta)^s_\Omega u = \lambda u$ we mean that

$$\frac{C_{N,s}}{2}\int_\Omega\int_\Omega \frac{(u(x)-u(y))(v(x)-v(y))}{|x-y|^{N+2s}}dxy = \lambda \int_\Omega uvdx, \ \forall\, v \in W_0^{s,2}(\Omega).$$

Let

$$W^{s-\frac{1}{2},2}(\partial\Omega) := \Big\{u|_{\partial\Omega}, \ u \in W^{s,2}(\Omega)\Big\}$$

be the trace space. Since $\lambda \in \mathbb{R}\backslash\sigma((-\Delta)^s_{\Omega,D})$, we have that the trace operator restricted to $\mathcal{H}^{s,\lambda}(\Omega)$, that is, the mapping $u \in \mathcal{H}^{s,\lambda}(\Omega) \mapsto u|_{\partial\Omega} \in W^{s-\frac{1}{2},2}(\partial\Omega)$, is linear and bijective. Letting

$$\|u|_{\partial\Omega}\|_{W^{s-\frac{1}{2},2}(\partial\Omega)} = \|u\|_{\mathcal{H}^{s,\lambda}(\Omega)},$$

then $W^{s-\frac{1}{2},2}(\partial\Omega)$ becomes a Hilbert space. By the closed graph theorem, different choice of $\lambda \in \mathbb{R}\backslash\sigma((-\Delta)^s_{\Omega,D})$ leads to an equivalent norm on $W^{s-\frac{1}{2},2}(\partial\Omega)$. Moreover, $W^{s-\frac{1}{2},2}(\partial\Omega) \overset{d}{\hookrightarrow} L^2(\partial\Omega)$ and one has the continuous embedding

$$W^{s-\frac{1}{2},2}(\partial\Omega) \hookrightarrow L^q(\partial\Omega) \tag{2.3.40}$$

with

$$1 \le q \le \frac{2(N-1)}{N-2s} \ \text{ if } N > 2s \text{ and } 1 \le q < \infty \text{ if } N \le 2s. \tag{2.3.41}$$

By [62], $\mathbb{D}_{s,\lambda}$ is the self-adjoint operator associated with the closed and symmetric form $\mathcal{E}_{s,\lambda}$ with domain $W^{s-\frac{1}{2},2}(\partial\Omega)$ and given by

$$\mathcal{E}_{s,\lambda}(\varphi,\psi) = \frac{C_{N,s}}{2}\int_\Omega\int_\Omega \frac{(u(x)-u(y))(v(x)-v(y))}{|x-y|^{N+2s}}dxdy - \lambda\int_\Omega uvdx,$$

where $\varphi, \psi \in W^{s-\frac{1}{2},2}(\partial\Omega)$ and $u, v \in \mathcal{H}^{s,\lambda}(\Omega)$ are such that $u|_{\partial\Omega} = \varphi$ and $v|_{\partial\Omega} = \psi$, and the operator $-\mathbb{D}_{s,\lambda}$ generates a strongly continuous and analytic semigroup on $L^2(\partial\Omega)$ which is also submarkovian if $\lambda \leq 0$. If $\lambda < 0$, then $\mathcal{E}_{s,\lambda}$ is also coercive. In that case we have that $\mathcal{E}_{s,\lambda}$ satisfies (2.2.9) with $\mathcal{X} = \partial\Omega$ and the constant $\mu = \frac{2(N-1)}{2s-1}$ if $N > 2s$ and $\mu > 2$ arbitrary if $N \leq 2s$. This implies that Propositions 2.2.1 and 2.2.2 hold with $A = -\mathbb{D}_{s,\lambda}$ in which case $\beta_A = \frac{N-1}{2s-1}$ if $N > 2s$ and $\beta_A > 1$ arbitrary if $N \leq 2s$. There is no regularity results available in the literature that can help to characterize the space $\mathcal{L}^\infty(\partial\Omega)$. We refer to [62, 65] for more details on this topic.

We conclude this section by considering fractional powers of operators. Before giving some concrete examples we introduce a general abstract theory. Let V and H be Hilbert spaces such that $V \overset{d}{\hookrightarrow} H$. Recall that in that case we have the Gelfand triple

$$V \overset{d}{\hookrightarrow} H \overset{d}{\hookrightarrow} V^\star.$$

Let $\mathcal{E} : V \times V \to \mathbb{R}$ be a symmetric, bilinear, continuous and coercive form. Let $\mathcal{A} : V \to V^\star$ be the operator given by $\langle \mathcal{A}u, v\rangle_{V^\star,V} = \mathcal{E}(u, v)$ for all $u, v \in V$.

Next we consider functions defined from $(0, \infty)$ into X where $X = V, H$ or $V^\star$. In general if X and Y are Hilbert spaces such that $X \overset{d}{\hookrightarrow} Y$ and $0 < s < 1$, we define the space

$$W_s(X, Y) := \Big\{\mathcal{U} \in L^1_{\text{loc}}((0, \infty); Y) : \mathcal{U}' \in L^1_{\text{loc}}((0, \infty); X),$$
$$\Big(t \mapsto t^s\mathcal{U}(t)\Big) \in L^2\left((0, \infty); Y, \frac{dt}{t}\right) \text{ and } \Big(t \mapsto t^s\mathcal{U}'(t)\Big) \in L^2\left((0, \infty); X, \frac{dt}{t}\right)\Big\}.$$

In order to avoid clutter we write t^s for the function $t \mapsto t^s$. It is clear that $W_s(X, Y)$ endowed with the norm

$$\|\mathcal{U}\|_{W_s(X,Y)} = \left(\int_0^\infty \Big(\|\mathcal{U}(t)\|_Y^2 + \|\mathcal{U}'(t)\|_X^2\Big)\, t^{2s-1}dt\right)^{\frac{1}{2}}$$

is a Hilbert space.

An s-**harmonic function with respect to** $\mathcal{E}$ is a function $\mathcal{U} \in W_{1-s}(H, V)$ such that $t^{1-2s}\mathcal{U}' \in W_s(V^\star, H)$ and

$$-\left(t^{1-2s}\mathcal{U}'\right)' + t^{1-2s}\mathcal{A}\mathcal{U} = 0 \text{ in } \mathcal{V}^\star \text{ for a.e. } t \in (0, \infty). \tag{2.3.42}$$

We notice that (2.3.42) is equivalent to the following equation

$$\mathcal{U}''(t) + \frac{1-2s}{t}\mathcal{U}'(t) - \mathcal{A}\mathcal{U}(t) = 0, \quad t \in (0, \infty).$$

It has been shown in [13, Theorem 3.4], that given $u \in [H, V]_s$ (the complex interpolation space), there exists a unique s-harmonic function $\mathcal{U}$ such that $\mathcal{U}(0) = u$. In addition, given $v \in [V^\star, H]_{1-s} = ([H, V]_s)^\star$, there exists a unique s-harmonic function $\mathcal{U}$ such that $-\lim_{t\downarrow 0} t^{1-2s}\mathcal{U}'(t) = v$. Now define the operator $\mathcal{D}_s : [H, V]_s \to ([H, V]_s)^\star$ as follows. Let $u \in [H, V]_s$ and let $\mathcal{U}$ be the unique s-harmonic function satisfying $\mathcal{U}(0) = u$. Then we set $\mathcal{D}_s u = v$ where $v := -\lim_{t\downarrow 0} t^{1-2s}\mathcal{U}'(t)$ in $V^\star$. We call $\mathcal{D}_s$ the **Dirichlet-to-Neumann operator with respect to $\mathcal{E}$**. Notice that

$$[H, V]_s \overset{d}{\hookrightarrow} H \overset{d}{\hookrightarrow} ([H, V]_s)^\star.$$

Now let D_s be the part of the operator $\mathcal{D}_s$ in H, that is,

$$D(D_s) = \Big\{u \in [H, V]_s : \mathcal{D}_s u \in H\Big\}, \quad D_s u = \mathcal{D}_s u.$$

Let also $\mathbb{A}$ be the part of the operator $\mathcal{A}$ in H and denote by $\mathbb{A}^s$ the fractional power of $\mathbb{A}$. By [13, Theorem 4.1] we have that $c_s D_s = \mathbb{A}^s$ where $c_s := 2^{2s-1}\frac{\Gamma(s)}{\Gamma(1-s)}$. In addition the operator $-\mathbb{A}^s$ generates a strongly continuous semigroup on H.

In the case where $H = L^2(\mathbb{R}^N)$, $V = W^{1,2}(\mathbb{R}^N)$,

$$\mathcal{E}(u, v) = \int_{\mathbb{R}^N} \nabla u \cdot \nabla v dx,$$

that is $\mathcal{A} = -\Delta$ (the Laplace operator on the whole space $\mathbb{R}^N$), the above construction of $\mathbb{A}^s$ is known as the Caffarelli-Silvestre extension [20] and in that case one has that $\mathbb{A}^s = (-\Delta)^s$. The extension of this construction to the case where A is a self-adjoint operator with compact resolvent has been done first by Stinga and Torrea [58] and later by several other authors. The description given above is taken from [13] where general operators associated to sesquilinear and continuous (both coercive and non-coercive) forms have been considered.

Next, assume that $H = L^2(X)$, $\mathcal{H}^s(X) := [L^2(X), V]_s$ (the complex interpolation space) and that the strongly continuous semigroup generated by the associated operator $-\mathbb{A}^s$ is submarkovian and ultracontractive. The latter is equivalent to the continuous embedding $\mathcal{H}^s(X) := [L^2(X), V]_s \hookrightarrow L^q(X)$ for some $q > 2$. Then our Propositions (2.2.1) and (2.2.2) hold in this abstract setting. Now we give some concrete examples.

Example 2.3.8 Throughout this example we assume that $\Omega \subset \mathbb{R}^N$ is a bounded open set with Lipschitz continuous boundary $\partial\Omega$.

(a) **The spectral fractional Dirichlet operator**. Let L be the uniformly elliptic operator introduced in (2.3.17). Let $\mathcal{E}_{L,D}$ with $D(\mathcal{E}_{L,D}) = W_0^{1,2}(\Omega)$ be given by

$$\mathcal{E}_{L,D}(u, v) = \sum_{i,j=1}^{N} \int_\Omega a_{ij}(x) D_i u D_j v dx.$$

Let $\mathcal{L}_D$ be the self-adjoint operator associated with $\mathcal{E}_{L,D}$ and denote by $(e^{-t\mathcal{L}_D})_{t\geq 0}$ the strongly continuous semigroup generated by $-\mathcal{L}_D$. Since $\mathcal{L}_D$ has a compact resolvent (this follows from the compact embedding $W_0^{1,2}(\Omega) \overset{c}{\hookrightarrow} L^2(\Omega)$) and $\mathcal{E}_{L,D}$ is coercive, we have that its eigenvalues form a non-decreasing sequence $0 < \lambda_1^D \leq \lambda_2^D \leq \cdots \leq \lambda_n^D \leq \cdots$ of real numbers satisfying $\lim_{n\to\infty} \lambda_n^D = \infty$. We denote by φ_n^D the orthonormal eigenfunction associated with λ_n^D.

For any $s \geq 0$, we also introduce the following fractional order Sobolev space:

$$\mathbb{H}_0^s(\Omega) := \left\{ u = \sum_{n=1}^{\infty} u_n \varphi_n^D \in L^2(\Omega) : \ \|u\|^2_{\mathbb{H}_0^s(\Omega)} := \sum_{n=1}^{\infty} (\lambda_n^D)^s u_n^2 < \infty \right\},$$

where $u_n = (u, \varphi_n^D)_{L^2(\Omega)} = \int_\Omega u\varphi_n^D \, dx$. If $0 < s < 1$, then

$$\mathbb{H}_0^s(\Omega) = \begin{cases} W^{s,2}(\Omega) = W_0^{s,2}(\Omega) & \text{if } 0 < s < \frac{1}{2}, \\ W_{00}^{\frac{1}{2},2}(\Omega) & \text{if } s = \frac{1}{2}, \\ W_0^{s,2}(\Omega) & \text{if } \frac{1}{2} < s < 1, \end{cases} \tag{2.3.43}$$

where

$$W_{00}^{\frac{1}{2},2}(\Omega) := \left\{ u \in W^{\frac{1}{2},2}(\Omega) : \ \int_\Omega \frac{|u(x)|^2}{(\mathrm{dist}(x, \partial\Omega))^2} dx < \infty \right\}.$$

In fact we have that

$$\mathbb{H}_0^s(\Omega) = \left[W_0^{1,2}(\Omega), L^2(\Omega) \right]_{1-s}.$$

It follows from (2.3.43) that (2.3.6) holds with $W_0^{s,2}(\Omega)$ replaced by $\mathbb{H}_0^s(\Omega)$.

Let $0 < s < 1$ and let $(\mathcal{L}_D)^s$ denote the fractional s-power of the operator $\mathcal{L}_D$. We describe three different ways to define $(\mathcal{L}_D)^s$. But all the three definitions coincide.

(i) The spectral fractional s power of $\mathcal{L}_D$ is defined on the space $\mathbb{H}_0^s(\Omega)$ by

$$(\mathcal{L}_D)^s u := \sum_{n=1}^{\infty} (\lambda_n^D)^s u_n \varphi_n^D \qquad \text{with } u_n = \int_\Omega u\varphi_n^D dx.$$

(ii) The operator $(\mathcal{L}_D)^s$ can be also defined by using the semigroup $(e^{-t\mathcal{L}_D})_{t\geq 0}$ as follows: for $u \in \mathbb{H}_0^s(\Omega)$ we set

$$(\mathcal{L}_D)^s u(x) = \frac{1}{\Gamma(-s)} \int_0^\infty \Big(e^{-t\mathcal{L}_D} u(x) - u(x)\Big) \frac{dt}{t^{1+s}}, \tag{2.3.44}$$

where $\Gamma(1-s) = -s\Gamma(-s)$.

(iii) Finally we have that $(\mathcal{L}_D)^s = c_s D_s$ where D_s is the Dirichlet-to-Neumann operator associated with $\mathcal{E}_{L,D}$, as constructed before the beginning of this example.

The operator $(\mathcal{L}_D)^s$ is unbounded, densely defined and with bounded inverse $(\mathcal{L}_D)^{-s}$ in $L^2(\Omega)$. But it can also be viewed as a bounded operator from $\mathbb{H}_0^s(\Omega)$ into its dual $(\mathbb{H}_0^s(\Omega))^\star$. The following integral representation of $(\mathcal{L}_D)^s$ given in [21, Theorem 2.3] will be useful. For every $u, v \in \mathbb{H}_0^s(\Omega)$, we have that

$$\begin{aligned} \langle(\mathcal{L}_D)^s u, v\rangle_{(\mathbb{H}_0^s(\Omega))^\star, \mathbb{H}_0^s(\Omega)} = & \int_\Omega\int_\Omega \Big(u(x) - u(y)\Big)\Big(v(x) - v(y)\Big) K_s(x, y) dx dy \\ & + \int_\Omega \kappa_s(x) u(x) v(x) dx, \end{aligned} \tag{2.3.45}$$

where

$$0 \leq K_s(x, y) := \frac{s}{\Gamma(1-s)} \int_0^\infty \frac{W_\Omega^D(t, x, y)}{t^{1+s}} dt, \quad x, y \in \Omega,$$

and

$$0 \leq \kappa_s(x) = \frac{s}{\Gamma(1-s)} \int_0^\infty \Big(1 - e^{-t\mathcal{L}_D} 1(x)\Big) \frac{dt}{t^{1+s}}, \quad x \in \Omega.$$

Here, W_Ω^D is the heat kernel associated to the semigroup $(e^{-t\mathcal{L}_D})_{t\geq 0}$, that is,

$$W_\Omega^D(t, x, y) = \sum_{n=1}^\infty e^{-t\lambda_n^D} \varphi_n^D(x) \varphi_n^D(y), \quad t > 0, \ x, y \in \Omega.$$

We mention that even if in the case $a_{i,j} = \delta_{ij}$, that is, $L = -\Delta$, the operator $(\mathcal{L}_D)^s$ is different from the operators $(-\Delta)_D^s$ and $(-\Delta)_{\Omega,D}^s$ introduced in Example 2.3.6 (a) and (d), respectively. For more details on this topic we refer to [57] and their references.

Firstly, we notice that it follows from (2.3.45) that $(\mathcal{L}_D)^s$ is associated with the closed, bilinear, symmetric, continuous and coercive form $\mathcal{E}_D^s$ with $D(\mathcal{E}_D^s) = \mathbb{H}_0^s(\Omega)$ and given by

$$\mathcal{E}_D^s(u,v) = \int_\Omega\int_\Omega \Big(u(x)-u(y)\Big)\Big(v(x)-v(y)\Big)K_s(x,y)\,dxdy + \int_\Omega \kappa_s(x)u(x)v(x)dx.$$

Secondly, let $u \in \mathbb{H}_0^s(\Omega)$. Proceeding as in [63, Lemma 2.7] we get that $u^+, u^- \in \mathbb{H}_0^s(\Omega)$ and $\mathcal{E}_D^s(u^+,u^-) \le 0$. This shows that the semigroup $(e^{-t(\mathcal{L}_D)^s})_{t\ge 0}$ generated by the operator $-(\mathcal{L}_D)^s$ is positive. Let $0 \le u \in \mathbb{H}_0^s(\Omega)$. Using [63, Lemma 2.7] again we get that $u \wedge 1 \in \mathbb{H}_0^s(\Omega)$ and $\mathcal{E}_D^s(u\wedge 1, u\wedge 1) \le \mathcal{E}_D^s(u,u)$. By Remark 2.2.6(ii) (see also e.g. [32, p.5]) this implies that the semigroup is submarkovian.

Thirdly, since $\mathbb{H}_0^s(\Omega)$ satisfies the embedding (2.3.6) with q given by (2.3.7), we have that the semigroup is ultracontractive.

In summary, we have shown that $\mathcal{E}_D^s$ satisfies (2.2.9) with $\mu = \frac{N}{s}$ if $N > 2s$ and $\mu > 2$ arbitrary if $N \le 2s$. Hence, Propositions 2.2.1 and 2.2.2 hold with $A = -(\mathcal{L}_D)^s$ in which case $\beta_A = \frac{N}{2s}$ if $N > 2s$ and $\beta_A > 1$ arbitrary if $N \le 2s$. In addition one can show that $\mathcal{L}^\infty(\Omega) = C_0(\Omega)$.

(b) **The spectral fractional Neumann operator**. Let L be as in (2.3.17) and let $b \in L^\infty(\Omega)$ be such that there exists a constant b_0 satisfying $b(x) \ge b_0 > 0$ a.e. on Ω. Let $\mathcal{E}_{L,\mathcal{N}}$ with $D(\mathcal{E}_{L,N}) = W^{1,2}(\Omega)$ be given by

$$\mathcal{E}_{L,\mathcal{N}}(u,v) = \sum_{i,j=1}^N \int_\Omega a_{ij}(x)D_iuD_jv\,dx + \int_\Omega b(x)uv\,dx.$$

Let $\mathcal{L}_\mathcal{N}$ be the self-adjoint operator associated with $\mathcal{E}_{L,\mathcal{N}}$ and denote by $(e^{-t\mathcal{L}_\mathcal{N}})_{t\ge 0}$ the strongly continuous semigroup generated by $-\mathcal{L}_\mathcal{N}$. Since $\mathcal{L}_\mathcal{N}$ has a compact resolvent (this follows from the compact embedding $W^{1,2}(\Omega) \overset{c}{\hookrightarrow} L^2(\Omega)$) and the form is coercive, it follows that its eigenvalues form a non-decreasing sequence $0 < \lambda_1^\mathcal{N} \le \lambda_2^\mathcal{N} \le \cdots \le \lambda_n^\mathcal{N} \le \cdots$ of real numbers satisfying $\lim_{n\to\infty}\lambda_n^\mathcal{N} = \infty$. Denoting by $\varphi_n^\mathcal{N}$ the orthonormal eigenfunction associated with $\lambda_n^\mathcal{N}$, then for $0 < s < 1$, the fractional s power $(\mathcal{L}_\mathcal{N})^s$ of the operator $\mathcal{L}_\mathcal{N}$ can be defined exactly as in parts (i), (ii) or (iii) above. In addition one also has the corresponding representation (2.3.45). Proceeding as in part (a) we get that Propositions 2.2.1 and 2.2.2 hold with $A = -(\mathcal{L}_\mathcal{N})^s$ in which case $\beta_A = \frac{N}{2s}$ if $N > 2s$ and $\beta_A > 1$ arbitrary if $N \le 2s$. In addition $\mathcal{L}^\infty(\Omega) = C(\overline{\Omega})$.

(c) **The spectral fractional Robin operator**. Let L be as in (2.3.17) and let $\beta \in L^\infty(\partial\Omega)$ satisfy $\beta(x) \geq \beta_0 > 0$ σ-a.e. on $\partial\Omega$, for some constant β_0. Let $\mathcal{E}_{L,R}$ with $D(\mathcal{E}_{L,R}) = W^{1,2}(\Omega)$ be given by

$$\mathcal{E}_{L,R}(u,v) = \sum_{i,j=1}^{N} \int_\Omega a_{ij}(x) D_i u D_j v dx + \int_{\partial\Omega} \beta(x) u v d\sigma.$$

Let $\mathcal{L}_R$ be the self-adjoint operator associated with $\mathcal{E}_{L,R}$ and denote by $(e^{-t\mathcal{L}_R})_{t\geq 0}$ the strongly continuous semigroup generated by $-\mathcal{L}_R$. Since $\mathcal{L}_R$ has a compact resolvent (this follows from the compact embedding $W^{1,2}(\Omega) \overset{c}{\hookrightarrow} L^2(\Omega)$) and the form is coercive, we have that its eigenvalues form a non-decreasing sequence $0 < \lambda_1^R \leq \lambda_2^R \leq \cdots \leq \lambda_n^R \leq \cdots$ of real numbers satisfying $\lim_{n\to\infty} \lambda_n^R = \infty$. Denote by φ_n^R the orthonormal eigenfunction associated with λ_n^R. Then for $0 < s < 1$, we define the fractional s power $(\mathcal{L}_R)^s$ of $\mathcal{L}_R$ as in parts (i), (ii) or (iii) above. In addition one also has the corresponding representation (2.3.45). Here also, proceeding as in part (a) we get that Propositions 2.2.1 and 2.2.2 hold with $A = -(\mathcal{L}_R)^s$ in which case $\beta_A = \frac{N}{2s}$ if $N > 2s$ and $\beta_A > 1$ arbitrary if $N \leq 2s$. In addition we have that $\mathcal{L}^\infty(\Omega) = C(\overline{\Omega})$.

We conclude this section with the following remark.

Remark 2.3.9 We mention the following facts.

(a) In our definition of the ultracontractivity, we have assumed that the estimate (2.2.3) holds for every $t > 0$. Usually if the estimate also holds for all $0 < t \leq 1$, or more precisely,

$$\|S(t)\|_{\mathcal{L}(L^p(\mathcal{X}),L^q(\mathcal{X}))} \leq C(t \wedge 1)^{-\beta_A\left(\frac{1}{p}-\frac{1}{q}\right)}, \ \forall t > 0, \tag{2.3.46}$$

then the semigroup is also ultracontractive. As we have seen in Theorem 2.2.5, the estimate (2.2.3) implies that the associated bilinear form is in particular coercive. Instead, (2.3.46) does not requires the form to be coercive.

(b) In all the above examples the bilinear forms associated with the considered operators are coercive. This is due to the fact that we would like to have the estimate (2.2.3). But we notice that this is not a restriction. Recall that we would like to investigate the existence and regularity of solutions to the system (1.0.1). If the bilinear form associated with the operator A is not coercive, then we write our system as

$$\begin{cases} \partial_t^\alpha u = Au - u + f(x,t,u) + u = \widetilde{A}u + \widetilde{f}(x,t,u) & \text{in } \mathcal{X} \times (0,T], \\ u(\cdot,0) = u_0 & \text{in } \mathcal{X}, \end{cases}$$

for some given $T > 0$, where $\widetilde{A}u = Au - u$ and $\widetilde{f}(x, t, u) = f(x, t, u) + u$. In that case the bilinear form associated with the operator $\widetilde{A}$ will be coercive and our new nonlinearity $\widetilde{f}$ will also satisfy the same assumptions as f.

From the above observations one can also consider that the semigroup S satisfies the estimate (2.3.46) which is more general than (2.2.3). Also in all the above Neumann type boundary conditions one does not need to consider a perturbation of the classical operator with the identity mapping.

References

1. N. Abatangelo, Large S-harmonic functions and boundary blow-up solutions for the fractional Laplacian. Discrete Contin. Dyn. Syst. **35**(12), 5555–5607 (2015)
2. D.R. Adams, L.I. Hedberg, *Function Spaces and Potential Theory*. Grundlehren der Mathematischen Wissenschaften [Fundamental Principles of Mathematical Sciences], vol. 314 (Springer, Berlin, 1996)
3. H. Antil, R. Khatri, M. Warma, External optimal control of nonlocal PDEs. Inverse Probl. **35**(8), 084003, 35 (2019)
4. D. Araya, C. Lizama, Almost automorphic mild solutions to fractional differential equations. Nonlinear Anal. **69**(11), 3692–3705 (2008)
5. W. Arendt, D. Daners, The Dirichlet problem by variational methods. Bull. Lond. Math. Soc. **40**(1), 51–56 (2008)
6. W. Arendt, D. Daners, Varying domains: stability of the Dirichlet and the Poisson problem. Discrete Contin. Dyn. Syst. **21**(1), 21–39 (2008)
7. W. Arendt, R. Mazzeo, Friedlander's eigenvalue inequalities and the Dirichlet-to-Neumann semigroup. Commun. Pure Appl. Anal. **11**(6), 2201–2212 (2012)
8. W. Arendt, A.F.M. ter Elst, The Dirichlet-to-Neumann operator on rough domains. J. Differ. Equ. **251**(8), 2100–2124 (2011)
9. W. Arendt, M. Warma, Dirichlet and Neumann boundary conditions: what is in between? J. Evol. Equ. **3**(1), 119–135 (2003). Dedicated to Philippe Bénilan
10. W. Arendt, M. Warma, The Laplacian with Robin boundary conditions on arbitrary domains. Potential Anal. **19**(4), 341–363 (2003)
11. W. Arendt, C.J.K. Batty, M. Hieber, F. Neubrander, *Vector-Valued Laplace Transforms and Cauchy Problems*. Monographs in Mathematics, vol. 96, 2nd edn. (Birkhäuser/Springer Basel AG, Basel, 2011)
12. W. Arendt, A.F.M. ter Elst, J.B. Kennedy, M. Sauter, The Dirichlet-to-Neumann operator via hidden compactness. J. Funct. Anal. **266**(3), 1757–1786 (2014)
13. W. Arendt, A.F.M. ter Elst, M. Warma, Fractional powers of sectorial operators via the Dirichlet-to-Neumann operator. Commun. Partial Differ. Equ. **43**(1), 1–24 (2018)
14. E.G. Bajlekova, *Fractional Evolution Equations in Banach Spaces*. Technische Universiteit Eindhoven, Eindhoven (2001)
15. U. Biccari, M. Warma, E. Zuazua, Local elliptic regularity for the Dirichlet fractional Laplacian. Adv. Nonlinear Stud. **17**(2), 387–409 (2017)
16. M. Biegert, D. Daners, Local and global uniform convergence for elliptic problems on varying domains. J. Differ. Equ. **223**(1), 1–32 (2006)
17. M. Biegert, M. Warma, The heat equation with nonlinear generalized Robin boundary conditions. J. Differ. Equ. **247**(7), 1949–1979 (2009)
18. M. Biegert, M. Warma, Some quasi-linear elliptic equations with inhomogeneous generalized Robin boundary conditions on "bad" domains. Adv. Differ. Equ. **15**(9–10), 893–924 (2010)

19. K. Bogdan, K. Burdzy, Z.-Q. Chen, Censored stable processes. Probab. Theory Relat. Fields **127**(1), 89–152 (2003)
20. L.A. Caffarelli, L. Silvestre, An extension problem related to the fractional Laplacian. Commun. Partial Differ. Equ. **32**(7–9), 1245–1260 (2007)
21. L.A. Caffarelli, P.R. Stinga, Fractional elliptic equations, Caccioppoli estimates and regularity. Ann. Inst. H. Poincaré Anal. Non Linéaire **33**(3), 767–807 (2016)
22. M. Caputo, Linear models of dissipation whose Q is almost frequency independent. II. Fract. Calc. Appl. Anal. **11**(1), 4–14 (2008). Reprinted from Geophys. J. R. Astr. Soc. **13**(5), 529–539 (1967)
23. B. Claus, M. Warma, Realization of the fractional laplacian with nonlocal exterior conditions via forms method. J. Evol. Equ. (2020). https://doi.org/10.1007/s00028-020-00567-0
24. Ph. Clément, S.-O. Londen, G. Simonett, Quasilinear evolutionary equations and continuous interpolation spaces. J. Differ. Equ. **196**(2), 418–447 (2004)
25. E. Cuesta, Asymptotic behaviour of the solutions of fractional integro-differential equations and some time discretizations. Discrete Contin. Dyn. Syst. 277–285 (2007). https://doi.org/10.3934/proc.2007.2007.277
26. D. Daners, Robin boundary value problems on arbitrary domains. Trans. Am. Math. Soc. **352**(9), 4207–4236 (2000)
27. E.B. Davies, *Heat Kernels and Spectral Theory*. Cambridge Tracts in Mathematics, vol. 92 (Cambridge University Press, Cambridge, 1990)
28. P.N.C. De Mendes, *Fractional Differential Equations: A Novel Study of Local and Global Solutions in Banach Spaces*. PhD thesis, Universidade de São Paulo, 2013
29. E. Di Nezza, G. Palatucci, E. Valdinoci, Hitchhiker's guide to the fractional Sobolev spaces. Bull. Sci. Math. **136**(5), 521–573 (2012)
30. S. Dipierro, X. Ros-Oton, E. Valdinoci, Nonlocal problems with Neumann boundary conditions. Rev. Mat. Iberoam. **33**(2), 377–416 (2017)
31. P. Felmer, A. Quaas, Boundary blow up solutions for fractional elliptic equations. Asymptot. Anal. **78**(3), 123–144 (2012)
32. M. Fukushima, Y. Oshima, M. Takeda, *Dirichlet Forms and Symmetric Markov Processes*. De Gruyter Studies in Mathematics, vol. 19, extended edition. (Walter de Gruyter & Co., Berlin, 2011)
33. C.G. Gal, M. Warma, Elliptic and parabolic equations with fractional diffusion and dynamic boundary conditions. Evol. Equ. Control Theory **5**(1), 61–103 (2016)
34. C.G. Gal, M. Warma, Long-term behavior of reaction-diffusion equations with nonlocal boundary conditions on rough domains. Z. Angew. Math. Phys. **67**(4), Art. 83, 42 (2016)
35. C.G. Gal, M. Warma, Reaction-diffusion equations with fractional diffusion on non-smooth domains with various boundary conditions. Discrete Contin. Dyn. Syst. **36**(3), 1279–1319 (2016)
36. R. Gorenflo, F. Mainardi, Fractional calculus: integral and differential equations of fractional order, in *Fractals and Fractional Calculus in Continuum Mechanics (Udine, 1996)*. CISM Courses and Lect., vol. 378 (Springer, Vienna, 1997), pp. 223–276
37. R. Gorenflo, Y. Luchko, F. Mainardi, Analytical properties and applications of the Wright function. Fract. Calc. Appl. Anal. **2**(4), 383–414 (1999). TMSF, AUBG'99, Part A (Blagoevgrad)
38. P. Grisvard, *Elliptic Problems in Nonsmooth Domains*. Classics in Applied Mathematics, vol. 69 (Society for Industrial and Applied Mathematics (SIAM), Philadelphia, 2011). Reprint of the 1985 original [MR0775683], With a foreword by Susanne C. Brenner
39. Q.-Y. Guan, Z.-M. Ma, Boundary problems for fractional Laplacians. Stoch. Dyn. **5**(3), 385–424 (2005)
40. Q.-Y. Guan, Z.-M. Ma, Reflected symmetric α-stable processes and regional fractional Laplacian. Probab. Theory Relat. Fields **134**(4), 649–694 (2006)
41. D. Henry, *Geometric Theory of Semilinear Parabolic Equations*. Lecture Notes in Mathematics, vol. 840 (Springer, Berlin, 1981)

42. V. Keyantuo, C. Lizama, M. Warma, Asymptotic behavior of fractional-order semilinear evolution equations. Differ. Integr. Equ. **26**(7–8), 757–780 (2013)
43. V. Keyantuo, C. Lizama, M. Warma, Spectral criteria for solvability of boundary value problems and positivity of solutions of time-fractional differential equations. Abstr. Appl. Anal. Art. ID 614328, 11 (2013). http://dx.doi.org/10.1155/2013/614328
44. V. Keyantuo, C. Lizama, M. Warma, Existence, regularity and representation of solutions of time fractional diffusion equations. Adv. Differ. Equ. **21**(9–10), 837–886 (2016)
45. L. Li, J.-G. Liu, Some compactness criteria for weak solutions of time fractional PDEs. SIAM J. Math. Anal. **50**, 3693–3995 (2018)
46. L. Li, J.-G. Liu, A generalized definition of caputo derivatives and its application to fractional odes. SIAM J. Math. Anal. **50**, 2867–2900 (2018)
47. C. Lizama, Regularized solutions for abstract Volterra equations. J. Math. Anal. Appl. **243**(2), 278–292 (2000)
48. F. Mainardi, Fractional calculus: some basic problems in continuum and statistical mechanics, in *Fractals and Fractional Calculus in Continuum Mechanics (Udine, 1996)*. CISM Courses and Lect., vol. 378 (Springer, Vienna, 1997), pp. 291–348.
49. V.G. Maz'ja, *Sobolev Spaces*. Springer Series in Soviet Mathematics (Springer, Berlin, 1985). Translated from the Russian by T. O. Shaposhnikova
50. V.G. Maz'ya, S.V. Poborchi, *Differentiable Functions on Bad Domains* (World Scientific, River Edge, 1997)
51. El.-M. Ouhabaz, *Analysis of Heat Equations on Domains*. London Mathematical Society Monographs Series, vol. 31 (Princeton University Press, Princeton, 2005)
52. I. Podlubny, *Fractional Differential Equations*. Mathematics in Science and Engineering, vol. 198 (Academic, San Diego, 1999). An introduction to fractional derivatives, fractional differential equations, to methods of their solution and some of their applications
53. J. Prüss, *Evolutionary Integral Equations and Applications*. Modern Birkhäuser Classics (Birkhäuser/Springer Basel AG, Basel, 1993). [2012] reprint of the 1993 edition
54. X. Ros-Oton, J. Serra, The Dirichlet problem for the fractional Laplacian: regularity up to the boundary. J. Math. Pures Appl. (9) **101**(3), 275–302 (2014)
55. F. Rothe, *Global Solutions of Reaction-Diffusion Systems*. Lecture Notes in Mathematics, vol. 1072 (Springer, Berlin, 1984)
56. S.G. Samko, A.A. Kilbas, O.I. Marichev, *Fractional Integrals and Derivatives* (Gordon and Breach Science Publishers, Yverdon, 1993). Theory and applications, Edited and with a foreword by S.M. Nikolskiui, Translated from the 1987 Russian original, Revised by the authors
57. R. Servadei, E. Valdinoci, On the spectrum of two different fractional operators. Proc. R. Soc. Edinb. Sect. A **144**(4), 831–855 (2014)
58. P.R. Stinga, J.L. Torrea, Extension problem and Harnack's inequality for some fractional operators. Commun. Partial Differ. Equ. **35**(11), 2092–2122 (2010)
59. M. Stynes, Too much regularity may force too much uniqueness. Fract. Calc. Appl. Anal. **19**(6), 1554–1562 (2016)
60. R.-N. Wang, D.-H. Chen, T.-J. Xiao, Abstract fractional Cauchy problems with almost sectorial operators. J. Differ. Equ. **252**(1), 202–235 (2012)
61. M. Warma, The Robin and Wentzell-Robin Laplacians on Lipschitz domains. Semigroup Forum **73**(1), 10–30 (2006)
62. M. Warma, A fractional Dirichlet-to-Neumann operator on bounded Lipschitz domains. Commun. Pure Appl. Anal. **14**(5), 2043–2067 (2015)
63. M. Warma, The fractional relative capacity and the fractional Laplacian with Neumann and Robin boundary conditions on open sets. Potential Anal. **42**(2), 499–547 (2015)
64. M. Warma, The fractional Neumann and Robin type boundary conditions for the regional fractional p-Laplacian. Nonlinear Differ. Equ. Appl. **23**(1), Art. 1, 46 (2016)
65. M. Warma, On a fractional (s, p)-Dirichlet-to-Neumann operator on bounded Lipschitz domains. J. Elliptic Parabol. Equ. **4**(1), 223–269 (2018)

66. M. Warma, Approximate controllability from the exterior of space-time fractional diffusive equations. SIAM J. Control Optim. **57**(3), 2037–2063 (2019)
67. E.M. Wright, The generalized Bessel function of order greater than one. Q. J. Math. Oxford Ser. **11**, 36–48 (1940)
68. K. Yosida, *Functional Analysis*. Classics in Mathematics (Springer, Berlin, 1995). Reprint of the sixth (1980) edition

Chapter 3
The Semilinear Parabolic Problem

In the present chapter, we rely on the crucial results of Chap. 2 to develop well-posedness results in the same spirit of Rothe [22] where second order elliptic operators in divergence form have been considered for the classical parabolic problem ($\alpha = 1$). In fact, we aim to extend parts of the theory in [22] not only to the case $\alpha \in (0, 1)$ but also by considering a larger class of operators A, beyond the case of second order elliptic operators, as well as to impose more general assumptions on the nonlinearity.

3.1 Maximal Mild Solution Theory

To this end, we need to introduce some further notations and basic definitions. Let $T > 0$ be fixed but otherwise arbitrary, $p \in [1, \infty]$ and $\delta \in [0, \infty)$. We begin with defining the Banach space

$$E_{p,\delta,T} := \Big\{ u : \mathcal{X} \times (0, T] \to \mathbb{R} \text{ measurable and } u(\cdot, t) \in L^p(\mathcal{X}) \text{ for a.e } t \in (0, T] \\ \text{and } \|u\|_{E_{p,\delta,T}} = |||u|||_{p,\delta,T} := \sup_{t \in (0,T]} (t \wedge 1)^{\delta} \|u(\cdot, t)\|_{L^p(\mathcal{X})} < \infty \Big\}.$$

We also introduce the Banach space

$$L_{p_1,p_2,T} := \Big\{ u : \mathcal{X} \times (0, T] \to \mathbb{R} \text{ measurable}, \|u\|_{L_{p_1,p_2,T}} = \|u\|_{p_1,p_2,T} < \infty \Big\},$$

C. G. Gal, M. Warma, *Fractional-in-Time Semilinear Parabolic Equations and Applications*, Mathématiques et Applications 84,
https://doi.org/10.1007/978-3-030-45043-4_3

where

$$\|u\|_{p_1,p_2,T} = \sup_{t_1,t_2\in[0,T],0\leq t_2-t_1\leq 1} \left(\int_{t_1}^{t_2} \|u(\cdot,\tau)\|_{L^{p_1}(\mathcal{X})}^{p_2}\, d\tau \right)^{\frac{1}{p_2}},$$

for $p_1, p_2 \in [1,\infty)$, with the obvious modifications when $p_1 = p_2 = \infty$. Also denote

$$L_{p_1,p_2} = L_{p_1,p_2,\infty} \text{ and } \|\cdot\|_{p_1,p_2} = \|\cdot\|_{p_1,p_2,\infty}.$$

In this case, for $u \in D(A)$ we can conveniently rewrite problem (1.0.1) as follows:

$$\partial_t^\alpha u = Au + f(x,t,u) \text{ in } \mathcal{X} \times (0,T],\ u(\cdot,0) = u_0 \text{ in } \mathcal{X}. \tag{3.1.1}$$

Our main goal in this section is to state sufficiently general conditions on f for which we can infer the existence of properly-defined solutions for (3.1.1). Once again, let $T \in (0,\infty)$ and denote by I a time interval of the form $[0,T]$, $[0,T)$ or $[0,\infty)$.

Definition 3.1.1 By a mild solution of (3.1.1) on the interval I, we mean that the measurable function u has the following properties:

(a) $u(\cdot,t) \in L^1(\mathcal{X})$, for all $t \in I\backslash\{0\}$.
(b) $f(\cdot,t,u(\cdot,t)) \in L^1(\mathcal{X})$, for almost all $t \in I\backslash\{0\}$.
(c) $\int_0^t \|f(\cdot,\tau,u(\cdot,\tau))\|_{L^1(\mathcal{X})}\, d\tau < \infty$, for all $t \in I$.
(d) $u(\cdot,t) = S_\alpha(t)u_0 + \int_0^t P_\alpha(t-\tau) f(\cdot,\tau,u(\cdot,\tau))\, d\tau$, for all $t \in I\backslash\{0\}$, where the integral is an absolutely converging Bochner integral in the space $L^1(\mathcal{X})$.
(e) The initial datum u_0 is assumed in the following sense:

$$\lim_{t\to 0^+} \|u(\cdot,t) - u_0\|_{L^{p_0}(\mathcal{X})} = 0,$$

for $u_0 \in L^{p_0}(\mathcal{X})$, if $p_0 \in [1,\infty)$, and $u_0 \in \mathcal{L}^\infty(\mathcal{X}) := \overline{D(A_\infty)}^{L^\infty(\mathcal{X})}$ if $p_0 = \infty$.

For simpler notation we define the functions

$$u(t) : x \in \mathcal{X} \mapsto u(t)(x) = u(x,t) \in \mathbb{R}$$

and

$$\widetilde{f}(t,u(t)) : x \in \mathcal{X} \mapsto \widetilde{f}(t,u(t))(x) = f(x,t,u(x,t)) \in \mathbb{R};$$

denote the superposition operator

$$\widetilde{f} : (t, v) \in [0, \infty) \times L^1(\mathcal{X}) \mapsto \widetilde{f}(t, v) = f(\cdot, t, v(\cdot)) \in L^1(\mathcal{X})$$

on its natural domain of definition provided that the $L^1(\mathcal{X})$-norm of $f(\cdot, t, v(\cdot))$ is finite. Thus, after dropping the "$\sim$", condition (d) can be written more simply as

$$u(t) = S_\alpha(t) u_0 + \int_0^t P_\alpha(t - \tau) f(\tau, u(\tau)) d\tau. \tag{3.1.2}$$

Of course, in the case $\alpha = 1$, both operators $S_\alpha(t)$, $P_\alpha(t)$ in (3.1.2) are simply replaced by the semigroup $S(t)$.

Remark 3.1.2 We recall that both operators S_α, S are strongly continuous on $L^{p_0}(\mathcal{X})$, if $p_0 \in [1, \infty)$; the semigroup $S(t)$ is not strongly continuous on $L^\infty(\mathcal{X})$. However, by definition we also have that S is strongly continuous on $\mathcal{L}^\infty(\mathcal{X})$. For simplicity of notation, in what follows for $p_0 \in [1, \infty]$, we also denote

$$\mathcal{L}^{p_0}(\mathcal{X}) = L^{p_0}(\mathcal{X}) \ \textit{if} \ p_0 \in [1, \infty).$$

Thus for every $p_0 \in [1, \infty]$ and $u_0 \in \mathcal{L}^{p_0}(\mathcal{X})$, we have

$$\lim_{t \to 0^+} \| S_\alpha(t) u_0 - u_0 \|_{L^{p_0}(\mathcal{X})} = 0.$$

We observe that condition (e) of Definition 3.1.1 holds if and only if

$$\lim_{t \to 0^+} \| S_\alpha(t) u_0 - u(\cdot, t) \|_{L^{p_0}(\mathcal{X})} = 0. \tag{3.1.3}$$

However, in view of this remark (3.1.3), we can introduce a more general version of mild solutions in the case when $u_0 \in L^\infty(\mathcal{X})$. The difference between the following mild solution and the one introduced above in Definition 3.1.1 is only in what sense the initial datum is satisfied.

Definition 3.1.3 Let $u_0 \in L^\infty(\mathcal{X})$. If a mild solution satisfies all conditions (a)–(d) of Definition 3.1.1, and (e) is replaced by (3.1.3) in the case $p_0 = \infty$, then we call this solution a **quasi-mild solution** on the interval I. In other words, the initial datum u_0 for the nonlinear problem (3.1.1) is assumed to be "as good as" for the corresponding linear problem with $f \equiv 0$.

We aim to establish the existence of locally defined mild solutions under some suitable assumptions on the nonlinear function f. Let $\gamma \in [1, \infty)$, $q_1, q_2 \in [1, \infty]$ and a function $c = c(x, t) \geq 0$ such that $c \in L_{q_1, q_2}$. These assumptions are as follows.

(F1) $f(x,t,\cdot):\mathbb{R}\to\mathbb{R}$ is a measurable function such that

$$|f(x,t,\xi)|\le c(x,t)(1+|\xi|)^{\gamma}, \text{ for all } \xi\in\mathbb{R}, \text{ a.e. } (x,t)\in\mathcal{X}\times(0,\infty).$$

(F2) For all $\xi,\eta\in\mathbb{R}$, assume the local Lipschitz condition

$$|f(x,t,\xi)-f(x,t,\eta)|\le c(x,t)(1+|\xi|+|\eta|)^{\gamma-1}|\xi-\eta|, \text{ a.e. } (x,t)\in\mathcal{X}\times(0,\infty).$$

(F3) There exists a positive increasing function $Q:\mathbb{R}_+\to\mathbb{R}_+$ such that

$$|f(x,t,\xi)|\le c(x,t)\,Q(|\xi|), \text{ for all } \xi\in\mathbb{R}, \text{ a.e. } (x,t)\in\mathcal{X}\times(0,\infty).$$

(F4) For all $\xi,\eta\in\mathbb{R}$, assume the local Lipschitz condition

$$|f(x,t,\xi)-f(x,t,\eta)|\le c(x,t)\,Q(|\xi|+|\eta|)|\xi-\eta|, \text{ a.e. } (x,t)\in\mathcal{X}\times(0,\infty).$$

We notice that conditions **(F3)** and **(F4)** are more general alternatives to **(F1)** and **(F2)**, respectively, since precise growth conditions (as $|\xi|,|\eta|\to\infty$) are not imposed for the nonlinearity f. Our main assumption on the operator A is the following.

(HA) The closed operator A generates a strongly continuous semigroup $(S(t))_{t\ge0}$ on $L^2(\mathcal{X})$ that satisfies all the assumptions of Propositions 2.2.1 and 2.2.2.

In particular, assumption **(HA)** implies all the estimates of (2.2.12); these estimates become crucial in our subsequent analysis. We employ a contraction argument in the Banach space $E_{p,\delta,T}$, $p\in[1,\infty]$, $p\ge p_0$, (with a singularity at $t=0$) to construct a solution u locally in time. In what follows, let β_A be the constant mentioned in Proposition 2.2.2 and set

$$\mathfrak{n}:=\beta_A\alpha>0, \text{ where } \beta_A>1,\ 0<\alpha\le1. \tag{3.1.4}$$

Our first result is concerned with the existence of locally defined mild solutions under suitable assumptions on the parameters p_0, γ, q_1 and q_2.

Theorem 3.1.4 (Local Existence) Assume *(**HA**) and either one of the following.*

(a) *Assume* **(F1)–(F2)** *for some* $\gamma\in[1,\infty)$, $q_1\in[1,\infty]\cap(\beta_A,\infty]$, $q_2\in(1/\alpha,\infty]$ *and let* $u_0\in L^{p_0}(\mathcal{X})$, *for some* $p_0\in[1,\infty)$ *such that*

$$\frac{\mathfrak{n}}{q_1}+\frac{1}{q_2}+(\gamma-1)\frac{\mathfrak{n}}{p_0}<\alpha.$$

(b) *Assume* **(F3)–(F4)** *for some* $q_1\in[1,\infty]\cap(\beta_A,\infty]$, $q_2\in(1/\alpha,\infty]$ *such that*

$$\frac{\mathfrak{n}}{q_1}+\frac{1}{q_2}<\alpha$$

and let $u_0\in\mathcal{L}^\infty(\mathcal{X})\subset L^\infty(\mathcal{X})$.

(c) *Assume* **(F1)–(F2)** *for some* $\gamma \in (1, \infty)$, $q_1 \in [1, \infty] \cap (\beta_A, \infty]$, $q_2 \in (1/\alpha, \infty]$ *and let* $u_0 \in L^{p_0}(\mathcal{X})$, *for some* $p_0 \in (1, \infty)$, *satisfy*

$$\frac{\mathfrak{n}}{q_1} + \frac{1}{q_2} + (\gamma - 1)\frac{\mathfrak{n}}{p_0} = \alpha.$$

Then there exists a time $T > 0$ *(depending on* u_0*) such that the initial value problem* (3.1.1) *has a unique mild solution in the sense of Definition 3.1.1 on the interval* $[0, T]$.

The assertions of Theorem 3.1.4 will follow after we prove the following three lemmas.

Lemma 3.1.5 *Assume that hypothesis (a) of Theorem 3.1.4 is satisfied. Then the assertion of Theorem 3.1.4 holds. Furthermore,* $u \in E_{p,\delta,T}$, *for some* $p \geq p_0$ *and* u *is unique in this space.*

Proof The proof is developed using the crucial ultracontractivity estimates of Proposition 2.2.2 (see also (2.2.12)). In this proof and elsewhere, the constant $C > 0$ is independent of the times t, τ, T. We shall explicitly state its further dependence on other parameters whenever necessary. Let now $p \in [1, \infty]$ such that $p \geq p_0$ with $q_1 \in [1, \infty] \cap (\beta_A, \infty]$, $q_2 \in (1/\alpha, \infty]$, $\gamma \in [1, \infty)$ satisfy

$$\begin{cases} \dfrac{1}{q_1} + \dfrac{\gamma}{p} \leq 1, \\[2ex] \dfrac{1}{q_2} + \gamma\delta < 1, \\[2ex] \dfrac{1}{q_2} + \dfrac{\mathfrak{n}}{q_1} + (\gamma - 1)\left(\delta + \dfrac{\mathfrak{n}}{p}\right) + \varepsilon < \alpha, \\[2ex] 0 \leq \delta := \dfrac{\mathfrak{n}}{p_0} - \dfrac{\mathfrak{n}}{p} < \alpha, \end{cases} \tag{3.1.5}$$

for a sufficiently small $\varepsilon > 0$. *We note that if* $p_0 \geq \max(\beta_A, 1)$ *we automatically have* $\delta \in [0, \alpha)$, *for arbitrary* $p \in [p_0, \infty]$ *while for* $p_0 < \beta_A$ *this holds provided that* $p_0 \leq p < p_0/(1 - p_0/\beta_A)$. *Having the restriction* $\delta \in [0, \alpha)$ *is required only in the case* $\alpha \in (0, 1)$ due to (2.2.12). When $\alpha = 1$, *such a restriction can be eliminated and we require instead that* $\delta \geq 0$. The conditions (3.1.5) are then sought for such p. We also remark that the third condition of (3.1.5) is an immediate consequence of the main condition in the statement of the theorem (see part (a), from which we can infer the existence of such a small ε). The proof exploits a Picard iteration argument. To this end, let $T \in (0, 1)$ and fix an element $u_1 \in E_{p,\delta,T}$ which is otherwise arbitrary. We define a sequence

$$u_{m+1}(\cdot,t) = S_\alpha(t)\, u_0 + \int_0^t P_\alpha(t-\tau)\, f(\tau, u_m(\cdot,\tau))\, d\tau,\ t \in (0,T], \tag{3.1.6}$$

for all $m \in \mathbb{N}$. We first show by induction that $u_m \in E_{p,\delta,T}$, for all $m \in \mathbb{N}$. To this end, let $s_1 \in [1,\infty]$ be such that

$$\frac{1}{q_1} + \frac{\gamma}{p} \le \frac{1}{s_1} \quad \text{and} \quad \frac{\mathfrak{n}}{s_1} + \frac{1}{q_2} + (\gamma-1)\,\delta + \varepsilon < \alpha + \frac{\mathfrak{n}}{p}, \tag{3.1.7}$$

and suppose that $u_m \in E_{p,\delta,T}$ is already known. The bound **(F1)** and the Hölder inequality with exponents $(s_1, s_1/(s_1-1))$ yield

$$\begin{aligned} &(t\wedge 1)^\delta \, \|u_{m+1}(\cdot,t)\|_{L^p(\mathcal{X})} \\ &\le (t\wedge 1)^\delta \, \|S_\alpha(t)\, u_0\|_{L^p(\mathcal{X})} + (t\wedge 1)^\delta \int_0^t \|P_\alpha(t-\tau)\|_{p,s_1} \|f(\tau, u_m(\cdot,\tau))\|_{L^{s_1}(\mathcal{X})}\, d\tau \\ &\le (t\wedge 1)^\delta \, \|S_\alpha(t)\|_{p,p_0} \|u_0\|_{L^{p_0}(\mathcal{X})} \\ &\quad + (t\wedge 1)^\delta \int_0^t \|P_\alpha(t-\tau)\|_{p,s_1} \|c(\cdot,\tau)\|_{L^{q_1}(\mathcal{X})} (\tau\wedge 1)^{-\gamma\delta} \times \\ &\qquad \times \left[(\tau\wedge 1)^\delta \left(\|1 + |u_m(\cdot,\tau)|\|_{L^p(\mathcal{X})}\right)\right]^\gamma d\tau, \end{aligned} \tag{3.1.8}$$

where we have also used that $q_1 \ge s_1$ and $p \ge \gamma s_1$. The first term on the right-hand side of (3.1.8) can be estimated owing to (2.2.12) for $p \ge p_0$ and the definition of $\delta = \mathfrak{n}/p_0 - \mathfrak{n}/p$, $\mathfrak{n} := \beta_A \alpha$. For the second term we apply Lemma A.0.1 (see Appendix) with $r(\tau) \equiv \|c(\cdot,\tau)\|_{L^{q_1}(\mathcal{X})}$ (note that $p_{s_2}(r) = \|c\|_{q_1,q_2}$), $\theta := \gamma\delta$ and $s_2 = q_2 \in (1/\alpha, \infty]$, whose assumptions are satisfied, owing to (3.1.5)–(3.1.7), since

$$\frac{1}{q_2} + \gamma\delta < 1, \quad \frac{1}{q_2} + \gamma\delta + \varepsilon \le \alpha + \delta$$

and

$$\frac{\mathfrak{n}}{s_1} - \frac{\mathfrak{n}}{p} + 1 - \alpha < 1 - \frac{1}{q_2}, \qquad \frac{\mathfrak{n}}{s_1} + \frac{1}{q_2} + \gamma\delta + \varepsilon < \alpha + \frac{\mathfrak{n}}{p} + \delta.$$

In the space $E_{p,\delta,T}$, from (3.1.8) and using (2.2.12) we get

$$\begin{aligned} &|||u_{m+1}|||_{p,\delta,T} \\ &\le C\, \|u_0\|_{L^{p_0}(\mathcal{X})} \\ &\quad + \sup_{t\in(0,T]} \left[(t\wedge 1)^\delta \int_0^t \|P_\alpha(t-\tau)\|_{p,s_1} (\tau\wedge 1)^{-\gamma\delta} \|c(\cdot,\tau)\|_{L^{q_1}(\mathcal{X})}\, d\tau\right] \times \end{aligned} \tag{3.1.9}$$

$$\times \left|\left|\left|1 + |u_m|\right|\right|\right|_{p,\delta,T}^{\gamma}$$

$$\leq C\left(\|u_0\|_{L^{p_0}(\mathcal{X})} + (T \wedge 1)^{\varepsilon} \|c\|_{q_1,q_2} \left|\left|\left|1 + |u_m|\right|\right|\right|_{p,\delta,T}^{\gamma}\right),$$

for some constant $C > 0$ independent of t and u_m, u_0. Henceforth, $u_{m+1} \in E_{p,\delta,T}$ and the claim is proved. Analogously, exploiting the Lipschitz condition **(F2)**, the Hölder inequality together with the application of Lemma A.0.1 as above, we also find the uniform estimate

$$\begin{aligned} &|||u_{m+1} - u_m|||_{p,\delta,T} \qquad (3.1.10)\\ &\leq C\,(T \wedge 1)^{\varepsilon} \|c\|_{q_1,q_2} \left|\left|\left|1 + |u_m| + |u_{m-1}|\right|\right|\right|_{p,\delta,T}^{\gamma-1} |||u_m - u_{m-1}|||_{p,\delta,T}, \end{aligned}$$

for all $m \geq 2$ and $T \in (0, 1)$. Define $U := |||u_1|||_{p,\delta,T} + 2\,|||u_2 - u_1|||_{p,\delta,T}$ and choose a small enough time $T_* \in (0, 1]$ such that

$$C\,(T_* \wedge 1)^{\varepsilon}\,(1 + 2U)^{\gamma-1} \leq \frac{1}{2}. \qquad (3.1.11)$$

By induction, it follows from (3.1.10) and (3.1.11) that

$$\begin{cases} |||u_m|||_{p,\delta,T_*} \leq U, & \text{for all } m \geq 1,\\ |||u_{m+1} - u_m|||_{p,\delta,T_*} \leq \frac{1}{2}\,|||u_m - u_{m-1}|||_{p,\delta,T_*}, & \text{for all } m \geq 2. \end{cases} \qquad (3.1.12)$$

Thus, by iteration in (3.1.12), the sequence $\{u_m\}_{m\in\mathbb{N}}$ is Cauchy in the Banach space E_{p,δ,T_*}. Thus, it has a limit $u \in E_{p,\delta,T_*}$ such that

$$\lim_{m\to\infty} |||u_m - u|||_{p,\delta,T_*} = 0. \qquad (3.1.13)$$

It now remains to show that the limit u has all the required properties of Definition 3.1.1, (a)–(e) on the time interval $[0, T_*]$. Property (a) is immediate since $u \in E_{p,\delta,T_*}$. By ignoring the factor $\|P_\alpha\,(t-s)\|_{p,s_1}$ in (3.1.9), (b) and (c) follow from the estimate

$$\begin{aligned} \|f\|_{1,1,T_*} &\leq \int_0^{T_*} \|f\,(\tau, u\,(\cdot,\tau))\|_{L^{s_1}(\mathcal{X})}\,d\tau \qquad (3.1.14)\\ &\leq \int_0^{T_*} (\tau \wedge 1)^{-\gamma\delta} \|c\,(\cdot,\tau)\|_{L^{q_1}(\mathcal{X})} \left[(\tau \wedge 1)^{\delta} \left\|1 + |u\,(\cdot,\tau)|\right\|_{L^p(\mathcal{X})}\right]^{\gamma} d\tau\\ &\leq \frac{C\,(q_2)}{1 - \gamma\delta - \frac{1}{q_2}}\,(T_*)^{1-\gamma\delta-\frac{1}{q_2}} \|c\|_{q_1,q_2} \left|\left|\left|1 + |u|\right|\right|\right|_{p,\delta,T_*}^{\gamma}, \end{aligned}$$

owing to the bound **(F1)**, the Hölder inequality, the fact that $0 < T_* \leq 1$ and $0 < \frac{1}{q_2} + \gamma\delta < 1$. Similar reasoning, using the Lipschitz bound **(F2)**, the properties of $P_\alpha(t)$ and the Hölder inequality once more, yields for all $t \in (0, T_*]$,

$$\left\| \int_0^t P_\alpha (t-\tau) \Big(f(\tau, u_m(\cdot,\tau)) - f(\tau, u(\cdot,\tau)) \Big) d\tau \right\|_{L^1(\mathcal{X})} \tag{3.1.15}$$

$$\leq \int_0^t \| f(\tau, u_m(\cdot,\tau)) - f(u(\cdot,\tau)) \|_{L^{s_1}(\mathcal{X})} d\tau$$

$$\leq \int_0^t (\tau \wedge 1)^{-\gamma\delta} \| c(\cdot,\tau)\|_{L^{q_1}(\mathcal{X})} d\tau \, |||1 + |u_m| + |u|||_{p,\delta,T_*}^{\gamma-1} |||u_m - u|||_{p,\delta,T_*}$$

$$\leq \frac{C(q_2)}{1-\gamma\delta-\frac{1}{q_2}} (T_*)^{1-\gamma\delta-\frac{1}{q_2}} \|c\|_{q_1,q_2} |||1 + |u_m| + |u|||_{p,\delta,T_*}^{\gamma-1} |||u_m - u|||_{p,\delta,T_*}$$

which converges to zero as $m \to \infty$, by (3.1.13). Both (3.1.13) and (3.1.15) allow us to take the limit in $L^1(\mathcal{X})$-norm as $m \to \infty$ in the integral equation (3.1.6) in order to deduce the integral equation in Definition 3.1.1, (d). For the last property (e), by Remark 3.1.2 it suffices to check that

$$\lim_{t\to 0^+} \|u(\cdot,t) - S_\alpha(t)u_0\|_{L^{p_0}(\mathcal{X})} = 0.$$

To this end, let $s_0 \in [1, \infty)$ be such that

$$\frac{1}{q_2} + \gamma\delta \leq \frac{1}{s_0} \text{ and } \frac{\mathfrak{n}}{s_0} + \frac{1}{q_2} + \gamma\delta + \varepsilon < \alpha + \frac{\mathfrak{n}}{p_0},$$

for some $\varepsilon \in (0, \alpha - 1/q_2]$. The subsequent computation is similar to (3.1.9) but now we apply the statement of Lemma A.0.1 (see Appendix) with the choices $p := p_0$, $s_1 := s_0$, $s_2 := q_2 \in (1/\alpha, \infty]$, $\theta := \gamma\delta$, $\delta := 0$ and $\varepsilon := \varepsilon$ (note again that $r(\tau) \equiv \|c(\cdot,\tau)\|_{L^{q_1}(\mathcal{X})}$ and $p_{s_2}(r) = \|c\|_{q_1,q_2}$). Of course, if $s_0 \geq p_0$ is arbitrary we have once again that $\mathfrak{n}/s_0 - \mathfrak{n}/p_0 \in [0, 2\alpha)$ is trivially satisfied, while if $s_0 < p_0$ one may choose s_0 sufficiently close to $p_0 \in [1, \infty)$ such that $1/s_0 < 2/\beta_A + 1/p_0$. Note that the assumptions of Lemma A.0.1 are satisfied with the above choices of $\delta, s_1, s_2, p, \varepsilon, \theta$, since

$$\frac{1}{q_2} + \gamma\delta < 1, \quad \frac{1}{q_2} + \varepsilon + \gamma\delta \leq \alpha$$

and

$$\frac{\mathfrak{n}}{s_0} + \frac{1}{q_2} < \alpha + \frac{\mathfrak{n}}{p_0} \quad \text{and} \quad \frac{\mathfrak{n}}{s_0} + \frac{1}{q_2} + \varepsilon < \alpha + \frac{\mathfrak{n}}{p_0}.$$

Indeed, the bound **(F1)** and by virtue of Hölder's inequality, for all $t \in (0, T_*]$ we have

$$\begin{aligned} &\|u(\cdot,t) - S_\alpha u_0\|_{L^{p_0}(\mathcal{X})} && (3.1.16)\\ &\leq \left(\int_0^t \|P_\alpha(t-\tau)\|_{p_0,s_0} (\tau\wedge 1)^{-\gamma\delta} \|c(\cdot,\tau)\|_{L^{q_1}(\mathcal{X})} d\tau\right) |||1+|u||||^{\gamma}_{p,\delta,T_*}\\ &\leq C(t\wedge 1)^{\varepsilon} \|c\|_{q_1,q_2} |||1+|u||||^{\gamma}_{p,\delta,T_*}, \end{aligned}$$

which implies the desired assertion (e) of Definition 3.1.1.

The uniqueness of the mild solution follows from a similar computation which resembles (3.1.10). Indeed, let $T \in (0, T_*]$ and let $u_1, u_2 \in E_{p,\delta,T}$ be any two mild solutions of (3.1.1) corresponding to the same initial datum u_0. As in (3.1.10), we get

$$|||u_1 - u_2|||_{p,\delta,T} \leq C(T\wedge 1)^{\varepsilon} \|c\|_{q_1,q_2} |||1+|u_1|+|u_2||||^{\gamma-1}_{p,\delta,T} |||u_1-u_2|||_{p,\delta,T}\,. \tag{3.1.17}$$

for all $T \in (0, T_*]$. Hence, there exists a small time $\widehat{T} \in (0, T_*]$ such that $u_1(\cdot,t) \equiv u_2(\cdot,t)$ for $t \in [0, \widehat{T}]$ and uniqueness over the whole interval $[0, T_*]$ follows by a continuation argument (see Theorem 3.1.10 below for more details). The proof is finished. □

Next, we derive the corresponding result in the case (c) of Theorem 3.1.4.

Lemma 3.1.6 *Assume that the hypothesis (c) of Theorem 3.1.4 is satisfied for some $\gamma \in (1,\infty)$ and $p_0 \in (1,\infty)$. Then the assertion of Theorem 3.1.4 holds. The mild solution is unique in the space $E_{w,p,\delta,T} \subset E_{p,\delta,T}$, for some $p \geq p_0$ (see* (3.1.19) *below for the definition of $E_{w,p,\delta,T}$).*

Proof Choose a value $p \in (1,\infty]$ such that $p \geq p_0$ and $q_1 \in [1,\infty] \cap (\beta_A,\infty]$, $q_2 \in (1/\alpha,\infty]$, $\gamma \in [1,\infty)$ satisfy

$$\begin{cases} \dfrac{1}{q_1} + \dfrac{\gamma}{p} \leq 1,\\[2mm] \dfrac{1}{q_2} + \gamma\delta < 1,\\[2mm] \dfrac{1}{q_2} + \dfrac{\mathfrak{n}}{q_1} + (\gamma-1)(\delta + \dfrac{\mathfrak{n}}{p}) = \alpha,\\[2mm] 0 \leq \delta = \dfrac{\mathfrak{n}}{p_0} - \dfrac{\mathfrak{n}}{p} < \alpha. \end{cases} \tag{3.1.18}$$

We may apply the whole statement of Lemma A.0.2 (see Appendix) with the following choices $p := p_0$, $q := p$ and the (singleton) $\Pi := \{u_0\} \subset L^{p_0}(\mathcal{X})$. Consider the functions g, w constructed in Lemma A.0.2 and recall that $(w(t))^{-\delta} = g(t)(t \wedge 1)^{-\delta}$. The proof is in the same spirit of [22, Lemma 8] where in the proof of previous Lemma 3.1.5, we perform the uniform estimates in a new (weighted) Banach space $E_{w,p,\delta,T} \subset E_{p,\delta,T}$ given by

$$E_{w,p,\delta,T} := \left\{u \in E_{p,\delta,T},\ \|u\|_{w,p,\delta,T} := \sup_{t\in(0,T]} \left((w(t))^{\delta} \|u(\cdot,t)\|_{L^p(\mathcal{X})}\right) < \infty\right\}. \tag{3.1.19}$$

As we mentioned already, the proof is based on the same iteration argument performed for the sequence (3.1.6) taking place now in the space $E_{w,p,\delta,T}$. First, to show that $u_m \in E_{w,p,\delta,T}$ is well-defined for all $m \in \mathbb{N}$, we again apply an induction argument. Suppose that $u_1 \in E_{w,p,\delta,T}$ is arbitrary and assume that $u_m \in E_{w,p,\delta,T}$ is already proved. Next choose $s_1 \in [1, \infty]$ such that the equality

$$\frac{\mathfrak{n}}{q_1} + \frac{\gamma\mathfrak{n}}{p} = \frac{\mathfrak{n}}{s_1} = \alpha - \frac{1}{q_2} - (\gamma - 1)\delta + \frac{\mathfrak{n}}{p} \tag{3.1.20}$$

holds. The bound **(F1)**, the estimate (A.0.6) (see Lemma A.0.2),

$$\|S_\alpha(t)\|_{p,p_0} \leq Cw(t)^{-\delta} = Cg(t)(t \wedge 1)^{-\delta}$$

and the Hölder inequality give

$$\|u_{m+1}\|_{w,p,\delta,T} \leq C\left(\|u_0\|_{L^{p_0}(\mathcal{X})} + \varphi(T)\,\|1 + |u_m|\|^{\gamma}_{w,p,\delta,T}\right),$$

where

$$\varphi(T) := \sup_{t\in(0,T]} (w(t))^{\delta} \int_0^t \|P_\alpha(t-\tau)\|_{p,s_1} (w(\tau))^{-\gamma\delta} \|c(\cdot,\tau)\|_{L^{q_1}(\mathcal{X})}\, d\tau \tag{3.1.21}$$

$$= (g(T))^{\gamma-1} \sup_{t\in(0,T]} (t \wedge 1)^{\delta} \int_0^t \|P_\alpha(t-\tau)\|^{-\gamma\delta}_{p,s_1} (\tau \wedge 1)^{-\delta\gamma} \|c(\cdot,\tau)\|_{L^{q_1}(\mathcal{X})}\, d\tau.$$

The second factor on the right-hand side of (3.1.21) can be estimated by application of Lemma A.0.1 with the choices $p := p$, $s_1 := s_1$, $s_2 := q_2$, $\theta := \gamma\delta$, $\delta := \delta$ and $\varepsilon := 0$ (as well as $r(s) \equiv \|c(\cdot,s)\|_{L^{q_1}(\mathcal{X})}$, where $p_{s_2}(r) = \|c\|_{q_1,q_2}$). Then one has $\varphi(T) \leq C(g(T))^{\gamma-1}\|c\|_{q_1,q_2}$ and $\lim_{T\to 0}\varphi(T) = 0$. Henceforth, $u_{m+1} \in E_{w,p,\delta,T}$, for all $m \in \mathbb{N}$ and the claim is proved. The rest of the proof goes exactly as in the proof of Lemma 3.1.5. We briefly mention the (modified) estimates without giving the full details. In view of the Lipschitz condition **(F2)** and the Hölder inequality, we get

$$\begin{aligned}&||u_{m+1}-u_m||_{w,p,\delta,T}\\&\quad\leq\varphi(T)\,||1+|u_m|+|u_{m-1}|||_{w,p,\delta,T}^{\gamma-1}\,||u_m-u_{m-1}||_{w,p,\delta,T},\end{aligned}$$

for all $m\geq 2$. As usual, defining $U:=||u_1||_{w,p,\delta,T}+2\,||u_2-u_1||_{w,p,\delta,T}$ and choosing a small enough time $T_*\in(0,1]$ such that $C\varphi(T_*)(1+2U)^{\gamma-1}\,\|c\|_{q_1,q_2}\leq 1/2$, we obtain the analogue of (3.1.12) in the space E_{w,p,δ,T_*} instead of E_{p,δ,T_*}. Therefore, we deduce again the existence of a limit point $u\in E_{w,p,\delta,T_*}$ such that

$$\lim_{m\to\infty}||u_m-u||_{w,p,\delta,T_*}=0.$$

The estimates (3.1.14) and (3.1.15) concerning the nonlinearity f are proved exactly as in Lemma 3.1.5. The estimate (3.1.16) concerning the initial datum u_0 reads

$$\|u(\cdot,t)-S_\alpha(t)u_0\|_{L^{p_0}(\mathcal{X})}\leq C(g(t))^\gamma\,\|c\|_{q_1,q_2}\,||1+|u|||_{w,p,\delta,T_*}^\gamma,\qquad(3.1.22)$$

for all $t\in[0,T_*]$. We now recall that $\lim_{t\to0}g(t)=0$ by Lemma A.0.2 (b) (see Appendix). Thus, u is a mild solution of (3.1.1) in the sense of Definition 3.1.1 on the time interval $[0,T_*]$. The uniqueness of the mild solution in the space E_{w,p,δ,T_*} follows from an argument that is similar to the computation (3.1.17); we omit the details. □

Remark 3.1.7 If the set of initial data $\Pi\subset L^{p_0}(\mathcal{X})$ is bounded and the set

$$k(\Pi):=\left\{u(\cdot,t)\|u(\cdot,t)\|_{L^p(\mathcal{X})}^{-1}:\ u(\cdot,t)\in L^p(\mathcal{X}),\ t\in[0,T_{\max}),\ \|u(\cdot,t)\|_{L^p(\mathcal{X})}\neq0\right\}$$

is precompact in $L^p(\mathcal{X})$, then the function $w=w(t)$ as well as the numbers U, T_* can be chosen independent of $u_0\in\Pi$.

We conclude the proof of Theorem 3.1.4 by verifying the following statement.

Lemma 3.1.8 *Assume* **(F3)–(F4)** *for some* $q_1\in(\beta_A,\infty]\cap[1,\infty]$, $q_2\in(1/\alpha,\infty]$ *and let* $u_0\in\mathcal{L}^\infty(\mathcal{X})\subset L^\infty(\mathcal{X})$. *Then the assertion of Theorem 3.1.4 is satisfied. Furthermore,* $u\in E_{\infty,0,T}$ *and* u *is unique in this space.*

Proof In this case $\delta=0$, $p_0=p=\infty$; we choose some $\varepsilon\in(0,\alpha-1/q_2)$ such that

$$\frac{n}{q_1}+\frac{1}{q_2}+\varepsilon<\alpha\qquad(3.1.23)$$

and let $u_0\in\mathcal{L}^\infty(\mathcal{X})$. The proof exploits a Picard iteration argument for the sequence (3.1.6) in the space $E_{\infty,0,T}$, for some $T\leq1$. As in the previous lemma, begin with $T \in (0,1)$ and fix an element $u_1 \in E_{\infty,0,T}$ which is otherwise arbitrary. We first show by induction that $u_m \in E_{\infty,0,T}$, for all $m \in \mathbb{N}$. Assume that $u_m \in E_{\infty,0,T}$ is already known, and proceed with an estimate for u_{m+1}, using the bound **(F3)** and the contractivity estimates of Proposition 2.2.1. We deduce that

$$\|u_{m+1}(t)\|_{L^\infty(\mathcal{X})} \tag{3.1.24}$$
$$\leq \|u_0\|_{L^\infty(\mathcal{X})} + Q\left(|||u_m|||_{\infty,0,T}\right)\left(\int_0^t \|P_\alpha(t-\tau)\|_{\infty,q_1} \|c(\cdot,\tau)\|_{L^{q_1}(\mathcal{X})}\,d\tau\right)$$

since $\|S_\alpha(t)\|_\infty \leq 1$. Owing also to the fact that $\mathfrak{n}/q_1 < 2\alpha$ (i.e., $q_1 > \beta_A/2$), we infer by (2.2.12),

$$\left\|t^{1-\alpha} P_\alpha(t)\right\|_{\infty,q_1} \leq C(\alpha,q_1)\, t^{-\frac{\mathfrak{n}}{q_1}}, \text{ for all } t \in (0,T]. \tag{3.1.25}$$

We now apply Lemma A.0.1 with $r(\tau) = \|c(\cdot,\tau)\|_{L^{q_1}(\mathcal{X})}$, $p := \infty$, $s_1 := q_1$, $s_2 := q_2$, $\theta := 0$, $\delta := 0$, $\varepsilon := \varepsilon$. Assumptions of Lemma A.0.1 with these choices of $p, s_1, s_2, \theta, \delta, \varepsilon$ are satisfied owing to (3.1.23). Therefore, we get

$$|||u_{m+1}|||_{\infty,0,T} \leq C\left(\|u_0\|_{L^\infty(\mathcal{X})} + T^\varepsilon \|c\|_{q_1,q_2} Q\left(|||u_m|||_{\infty,0,T}\right)\right),$$

which proves that $u_{m+1} \in E_{\infty,0,T}$. By a similar argument, using the local Lipschitz condition **(F4)** and the Hölder inequality, we get that

$$|||u_{m+1} - u_m|||_{\infty,0,T} \tag{3.1.26}$$
$$\leq CT^\varepsilon \|c\|_{q_1,q_2} Q\left(|||u_m|||_{\infty,0,T} + |||u_{m-1}|||_{\infty,0,T}\right) |||u_m - u_{m-1}|||_{\infty,0,T},$$

for all $m \in \mathbb{N}\setminus\{1\}$. Let $U := |||u_1|||_{\infty,0,T} + 2\,|||u_2 - u_1|||_{\infty,0,T}$ and choose a small enough time $T_* \in (0,1]$ such that

$$CT_*^\varepsilon \|c\|_{q_1,q_2} Q(2U) \leq \frac{1}{2}.$$

It follows from (3.1.26) by induction that

$$\begin{cases} |||u_m|||_{\infty,0,T_*} \leq U, & \text{for all } m \geq 1, \\ |||u_{m+1} - u_m|||_{\infty,0,T_*} \leq \frac{1}{2}|||u_m - u_{m-1}|||_{\infty,0,T_*}, & \text{for all } m \geq 2. \end{cases} \tag{3.1.27}$$

Thus, by iteration in (3.1.27), the sequence $\{u_m\}_{m\in\mathbb{N}}$ is Cauchy in the Banach space $E_{\infty,0,T_*}$. Thus, it has a limit $u \in E_{\infty,0,T_*}$ such that

$$\lim_{m\to\infty} |||u_m - u|||_{\infty,0,T_*} = 0.$$

The latter convergence and the local Lipschitz condition **(F4)** then implies

$$\lim_{m\to\infty} |||f(\cdot, u_m) - f(\cdot, u)|||_{\infty,0,T_*} = 0.$$

Hence, as usual we can take the limit in the iteration (3.1.6) in the norm of $E_{\infty,0,T_*}$. More precisely, we see that u satisfies the integral equation and is indeed a mild solution on the interval $[0, T_*]$. For the uniqueness argument, we let $u_1, u_2 \in E_{\infty,0,T}$ be two mild solutions for any $T \in (0, T_*]$ with the same initial data u_0. Similarly to (3.1.26), we get

$$\begin{aligned}
&|||u_1 - u_2|||_{\infty,0,T} \\
&\le CT^{\varepsilon} \|c\|_{q_1,q_2} \mathcal{Q}\left(|||u_1|||_{\infty,0,T} + |||u_2|||_{\infty,0,T}\right) |||u_1 - u_2|||_{\infty,0,T},
\end{aligned}$$

and therefore, $u_1 \equiv u_2$ on $[0, T_0]$ for some $T_0 \in (0, T]$. Now uniqueness on the whole interval $[0, T_*]$ follows by a continuation argument (see Theorem 3.1.10 below for more details).

Finally, it remains to check that the solution satisfies the initial condition (see Definition 3.1.1, (e)). We let $u_0 \in \mathcal{L}^{\infty}(\mathcal{X})$ and estimate using the bound **(F3)** as follows:

$$\begin{aligned}
&\|u(\cdot, t) - S_\alpha(t)u_0\|_{L^\infty(\mathcal{X})} \\
&\le \mathcal{Q}\left(|||u|||_{\infty,0,T}\right)\left(\int_0^t \|P_\alpha(t-\tau)\|_{\infty,q_1} \|c(\cdot,\tau)\|_{L^{q_1}(\mathcal{X})}\, d\tau\right).
\end{aligned}$$

All the same arguments leading to the proof of (3.1.26), then give

$$\|u(\cdot, t) - S_\alpha(t)u_0\|_{L^\infty(\mathcal{X})} \le Ct^{\varepsilon} \|c\|_{q_1,q_2} \mathcal{Q}\left(|||u|||_{\infty,0,T}\right), \tag{3.1.28}$$

for all $t \in (0, T_*]$. This implies the desired claim in view of Remark 3.1.2, which then implies that

$$\lim_{t\to 0^+} \|u(\cdot, t) - u_0\|_{L^\infty(\mathcal{X})} = 0.$$

The proof of lemma is finished. □

We also conclude that the mild solution is also locally bounded in time in the space $L^{\infty}(\mathcal{X})$ for as long as it exists.

Theorem 3.1.9 (Local Boundedness) *Let the assumptions of Theorem 3.1.4 be satisfied. Then the mild solution of problem* (3.1.1) *on the interval* $[0, T]$ *satisfies*

$$\sup_{t\in[T_0,T]} \|u(\cdot, t)\|_{L^\infty(\mathcal{X})} < \infty, \text{ for all } T_0 \in (0, T]. \tag{3.1.29}$$

Proof It suffices to check (3.1.29) on $(0, T]$, where $T > 0$ is the existence time defined in the statement of Theorem 3.1.4. Obviously, the case $p_0 = \infty$ is already contained in the proof of Theorem 3.1.4 (part (b)), so it suffices to take $p_0 < \infty$. Consider the two sequences $\{p_i\}$, $\{\delta_i\}$ as constructed by Lemma A.0.3 (see Appendix) such that $\delta_i \in (0, \alpha)$, $i = 1, .., k$, and $p_0 < p_1 < \ldots < p_k = \infty$. One can now inductively apply for each $i = 1, .., k$, the statements of Lemma 3.1.5 (in the case (a)) and Lemma 3.1.6 (in the case (c)), to show

$$\sup_{t\in[T_0,T]} \|u(\cdot, t)\|_{L^{p_i}(\mathcal{X})} \leq C_{T_0,T}, \; i = 1, \ldots, k. \tag{3.1.30}$$

We start the argument by applying the statements of Lemmas 3.1.5, 3.1.6 with the following choices $p_0 := p_0$, $p := p_1$, $\delta := \delta_1$ and initial data $u_0 \in L^{p_0}(\mathcal{X})$. More precisely, there exist a time $T > 0$ and a mild solution u on $[0, T]$ in the sense of Definition 3.1.1. The mild solution satisfies (3.1.30) for $i = 1$ since $u \in E_{p_1,\delta_1,T}$. We now use an induction argument; assume that (3.1.30) is already known for some $i = j-1$ ($j \geq 2$) and prove that it also holds for $i = j$. Let $T_0 \in (0, T]$ be arbitrary. For all $\tau \in [T_0, T]$ we apply once again Lemma 3.1.5 with the choices $p_0 := p_{j-1}$, $p := p_j$, $\delta := \delta_i$ and use $u_0 := u(\tau)$ as an initial datum. Hence there exist a time $T_j \in (0, 1]$ (independent of τ since (3.1.30) holds with $i = j - 1$) and mild solutions u_j^τ associated with initial data $u(\tau)$ on the time interval $\left[\tau, \tau + T_j\right]$, for all $\tau \in [T_0, T]$. Furthermore, $u_j^\tau(\cdot, \tau + \cdot) \in E_{p_j,\delta_j,T_j}$ is also unique in this space. Hence, by uniqueness we infer that $u_j^\tau(t) = u(t)$ for all $t \in (\tau, \tau + T_i) \cap (0, T]$. Besides the conclusion of Lemma 3.1.5 (as well as Lemma 3.1.6, in the case (c) of Theorem 3.1.4) gives

$$\sup_{\tau\in[T_0,T]} \left|\left|\left|u_j^\tau\right|\right|\right|_{p_j,\delta_j,T_j} < \infty,$$

and since $T_0 \in (0, T]$ is chosen arbitrary, the latter implies the statement (3.1.30) for $i = j$. This concludes the induction argument. We can now apply it inductively to get (3.1.30) for $i = k$ (and so $p_k = \infty$), which is also the desired claim (3.1.29) of the theorem. □

We can now conclude with the result on the existence of (maximally-defined) mild solutions.

Theorem 3.1.10 (Existence of Maximal Mild Solutions) *Let the assumptions of Theorem 3.1.4 be satisfied. Then the mild solution of problem* (3.1.1) *has a maximal time interval of existence* $[0, T_{\max})$ *and either* $T_{\max} = \infty$, *or* $0 < T_{\max} < \infty$ *and*

$$\lim_{t\to T_{\max}} \|u(\cdot, t)\|_{L^{p_0}(\mathcal{X})} = \infty. \tag{3.1.31}$$

In the case (c) of Theorem 3.1.4, (3.1.31) *only holds under the additional assumption that the set*

$$\kappa(\Pi) := \Big\{ u(\cdot,t) \|u(\cdot,t)\|^{-1}_{L^{p_0}(\mathcal{X})} : u(\cdot,t) \in L^{p_0}(\mathcal{X}),\ and,$$
$$t \in [0, T_{\max}),\ \|u(\cdot,t)\|_{L^{p_0}(\mathcal{X})} \neq 0 \Big\}$$

is precompact in $L^{p_0}(\mathcal{X})$*. Finally, every mild solution satisfies*

$$\sup_{t\in[T_1,T_2]} \|u(\cdot,t)\|_{L^\infty(\mathcal{X})} < \infty, \text{ for all } T_1, T_2 \in (0, T_{\max}). \tag{3.1.32}$$

Proof A local mild solution on the interval $(0, T]$ for some initial datum $u_0 \in \mathcal{L}^{p_0}(\mathcal{X})$, $p_0 \in [1, \infty]$ was constructed in Theorem 3.1.4. This solution can be extended to a global one on $[0, T_{\max})$ with the aforementioned properties (3.1.31)–(3.1.32). This follows from a continuation argument, owing to the conclusions of Lemmas 3.1.5, 3.1.6, 3.1.8 and 3.1.9, and another constructive argument that is given below (see **Step I–II**). For the sake of convenience, we outline these arguments in the second case (b) of the theorem. In order to extend the local solution to a global one, one employs a simple inductive procedure. Similar procedures are applied also in the other two cases (a), (c) of Theorem 3.1.4.

To this end, define an increasing sequence $\{T_m\}$ and let u_m be the mild solution for the initial datum u_0 on the intervals $[0, T_m]$, for all $m \in \mathbb{N}$, defined as follows:

- Let $T_1 = T$ and u_1 be the local mild solution for initial datum u_0 on the interval $[0, T]$, furnished by the statement of Theorem 3.1.9.
- Assume that the mild solution u_m on the interval $[0, T_m]$ is already defined, as

$$u_m(t) = S_\alpha(t)u_0 + \int_0^t P_\alpha(t-s) f(s, u_m(s))\, ds,$$

  and apply the **Step II** provided below (at the end of the proof) to construct an extension v_m of u_m. Then there exist $T_m < T_{m+1}$ and a mild solution v_m for the problem on the interval $[0, T_{m+1}]$. The sequence of mild solutions u_{m+1} is naturally defined as

$$u_{m+1}(t) = \begin{cases} u_m(t), & t \in (0, T_m], \\ v_m(t), & t \in (T_m, T_{m+1}]. \end{cases}$$

  In particular, u_{m+1} is a mild solution of (2.1.6) on the time interval $[0, T_{m+1}]$. Clearly, $u_{m+1} \in E_{\infty,0,T_{m+1}}$ and $\|u_m(T_m)\|_{L^\infty(\mathcal{X})} \le U < \infty$, for $m \ge 1$ (see the proof of Lemma 3.1.8).
- The maximal time $T_{\max} > 0$ is defined as $\lim_{m\to\infty} T_m = T_{\max}$.

Step I. The mild solution $u(t)$ for the initial datum u_0 on the interval $[0, T_{\max})$ is defined as $u(t) = u_m(t)$, for all $t \in (0, T_m]$, $m \in \mathbb{N}$. Since by Lemma 3.1.8 and Theorem 3.1.9, we have

$$\sup_{t \in [T_m, T_{m+1}]} \|u(t)\|_{L^\infty(\mathcal{X})} < \infty, \quad \text{for all } m \in \mathbb{N}$$

and $\sup_{t \in [T_0, T_1]} \|u(t)\|_{L^\infty(\mathcal{X})} < \infty$, for all $T_0 \in (0, T_1]$, it follows that the mild solution u satisfies (3.1.32). It remains to prove that if $T_{\max} < \infty$, then

$$\limsup_{m \to \infty} \|u(t)\|_{L^\infty(\mathcal{X})} = \infty. \tag{3.1.33}$$

One argues by contradiction by assuming that if (3.1.33) does not hold, then $T_{\max} < \infty$ and

$$\lim_{t \in [T_1, T_{\max})} \|u(t)\|_{L^\infty(\mathcal{X})} = U < \infty. \tag{3.1.34}$$

As one can see from the arguments provided below in **Step II**, the length of the interval $[T_m, T_{m+1}]$ of the extension v_m depends only on $q_1, q_2, \|c\|_{q_1, q_2}$, the function Q from the assumptions of the function f, and also non-increasingly in terms of $\|u_m(T_m)\|_{L^\infty(\mathcal{X})}$ (and so on U). Then (3.1.34) implies that $\sup_{m \in \mathbb{N}} (T_{m+1} - T_m) = \tau > 0$, which contradicts the fact that $\lim_{m \to \infty} T_m = T_{\max} < \infty$. Therefore, (3.1.33) must hold and the proof of **Step I** is finished.

Finally, in the final **Step II**, we provide the constructive details for the extension $v_m \in \mathbb{K}_\tau$, for some $\tau \in (0, 1]$. More precisely, the set $\mathbb{K}_\tau$ is defined as the set of all functions that satisfy $v_m(t) = u_m(t)$, for all $t \in [0, T_m]$, and such that

$$\|v_m(t) - u_m(T_m)\|_{L^\infty(\mathcal{X})} \le R, \quad \text{for all } t \in [T_m, T_m + \tau].$$

Next, we define the mapping

$$\Phi(v_m(t)) = S_\alpha(t) u_0 + \int_0^t P_\alpha(t - s) f(s, v_m(s))\, ds.$$

Our goal is to show that $\Phi : \mathbb{K}_\tau \to \mathbb{K}_\tau$ is a contraction, for some appropriately chosen $\tau, R > 0$. Notice that, for all $t \in [0, T_m]$, we have $\Phi(v_m(t)) = v_m(t)$, and so there is nothing to prove since u_m is already the unique fixed point of $\Phi(v_m) = \Phi(u_m)$. For $t \in [T_m, T_m + \tau]$, we first notice that

$$\begin{aligned}
&\Phi(v_m(t)) - u_m(T_m) \\
&= S_\alpha(t) u_0 - S_\alpha(T_m) u_0 + \int_{T_m}^t P_\alpha(t - s) f(s, v_m(s))\, ds \\
&= S_\alpha(t) u_0 - S_\alpha(T_m) u_0 + \int_0^{t - T_m} P_\alpha(t - T_m - s) f(s + T_m, v_m(s + T_m))\, ds.
\end{aligned}$$

It follows (owing to $\|S_\alpha(t)\|_\infty \leq 1$, for any $t \geq 0$, and estimates (3.1.24)–(3.1.25); also set $\|w\|_{\infty,\tau} = \sup_{t\in[T_m,T_m+\tau]} \|w(t)\|_{L^\infty(\mathcal{X})}$) that

$$\begin{aligned}
&\|\Phi(v_m(t)) - u_m(T_m)\|_{L^\infty(\mathcal{X})} \\
&\leq 2\|u_0\|_{L^\infty(\mathcal{X})} \\
&+ Q\left(\|v_m\|_{\infty,\tau}\right) \int_0^{t-T_m} \|P_\alpha(t-T_m-s)\|_{\infty,q_1} \|c(\cdot,s)\|_{L^{q_1}(\mathcal{X})} ds \\
&\leq 2\|u_0\|_{L^\infty(\mathcal{X})} + \tau^\varepsilon \|c\|_{q_1,q_2} Q\left(\|v_m\|_{\infty,\tau}\right) \\
&\leq 2\|u_0\|_{L^\infty(\mathcal{X})} + \tau^\varepsilon \|c\|_{q_1,q_2} Q(U+R),
\end{aligned}$$

since $\|u_m(T_m)\|_{L^\infty(\mathcal{X})} \leq U$. Choosing now $R \geq 4\|u_0\|_{L^\infty(\mathcal{X})}$ and a sufficiently small $\tau \in (0,1]$ such that

$$\tau^\varepsilon \|c\|_{q_1,q_2} Q(U+R) \leq \frac{R}{2}, \tag{3.1.35}$$

shows that Φ is well defined mapping from $\mathbb{K}_\tau$ to $\mathbb{K}_\tau$. A similar argument shows that, for any $v_m, w_m \in \mathbb{K}_\tau$, and $t \in [T_m, T_m+\tau]$,

$$\begin{aligned}
&\|\Phi(v_m(t)) - \Phi(w_m(t))\|_{L^\infty(\mathcal{X})} \\
&\leq Q\left(\|v_m\|_{\infty,\tau} + \|w_m\|_{\infty,\tau}\right) \|v_m - w_m\|_{\infty,\tau} \\
&\times \int_0^{t-T_m} \|P_\alpha(t-T_m-s)\|_{\infty,q_1} \|c(\cdot,s)\|_{L^{q_1}(\mathcal{X})} ds \\
&\leq C\tau^\varepsilon \|c\|_{q_1,q_2} Q(2R+2U) \|v_m - w_m\|_{\infty,\tau},
\end{aligned}$$

for some positive constant C independent of τ, T_m. Thus, Φ is a contraction on $\mathbb{K}_\tau$ provided that $\tau \in (0,1]$ satisfies (3.1.35) and

$$C\tau^\varepsilon \|c\|_{q_1,q_2} Q(2R+2U) \leq \frac{1}{2}.$$

Therefore, we may conclude that Φ has a unique fixed point $v_m \in \mathbb{K}_\tau$. This completes the proof of the final **Step II** (and, of the theorem). □

In addition, we can conclude with the following.

Corollary 3.1.11 *Let* $u_0 \in L^\infty(\mathcal{X})$ *and assume* **(F3)–(F4)** *for some* $q_1 \in (\beta_A, \infty] \cap [1, \infty]$, $q_2 \in (1/\alpha, \infty]$ *such that*

$$\frac{n}{q_1} + \frac{1}{q_2} < \alpha.$$

Then there exists a unique quasi-mild solution on $[0, T_{\max})$ *in the sense of Definition 3.1.3, such that either* $T_{\max} = \infty$*, or* $T_{\max} < \infty$ *and* (3.1.31) *is satisfied with* $p_0 = \infty$*. Furthermore, it holds* (3.1.32).

3.2 Maximal Strong Solution Theory

We are next concerned with further regularity properties for the mild solution of (3.1.1). To this end, we introduce the notion of strong solution for the semilinear problem (3.1.1).

Definition 3.2.1 Let $p \in (1, \infty)$ and $\alpha \in (0, 1]$. By a strong solution u of (3.1.1) on the time interval $I = (0, T)$ we mean

(a) u is a mild solution in the sense of Definition 3.1.3 *(with* $p_0 = \infty$*)*, where the initial datum $u(0) = u_0$ is meant in the following sense:

$$\lim_{t \to 0^+} \|u(\cdot, t) - u_0\|_{L^\infty(\mathcal{X})} = 0. \tag{3.2.1}$$

(b) $u \in C^{0,\kappa}(I; L^\infty(\mathcal{X}))$, for some $\kappa > 0$.
(c) $u(\cdot, t) \in D(A_p)$, for all $t \in I$ and $\partial_t^\alpha u \in C(I; L^p(\mathcal{X}))$.
(d) $\partial_t^\alpha u(\cdot, t) = A_p u(\cdot, t) + f(\cdot, t, u(\cdot, t))$ is satisfied for $t \in I$.

Theorem 3.2.2 (Strong Solutions on $(0, T_{\max})$**)** *Let* $q_1 \in (\beta_A, \infty] \cap [1, \infty]$, $0 < \alpha \leq 1$ *and* $u_0 \in D(A_p)$ *for some* $p \in (\beta_A, \infty) \cap (1, \infty)$*. Consider the following alternatives:*

(a) If $p \geq q_1$*, assume* f *obeys conditions* **(F3)–(F4)** *with* $q_2 \in (1/\alpha, \infty]$ *and* $\theta \in (\beta_A/p, 1)$ *satisfying*

$$\alpha(1 - \theta) - n\left(\frac{1}{q_1} - \frac{1}{p}\right) > \frac{1}{q_2}. \tag{3.2.2}$$

(b) If $p \leq q_1$*, assume* f *obeys conditions* **(F3)–(F4)** *with* $q_2 \in (1/\alpha, \infty]$ *and* $\theta \in (\beta_A/p, 1)$ *satisfying*

$$\alpha(1 - \theta) > \frac{1}{q_2}. \tag{3.2.3}$$

Suppose either (a) or (b) is satisfied and consider the following assumption:

(F5) *For all* $t, s > 0$, *for almost*[1] *all* $x \in \mathcal{X}$ *and* $\max\{|\xi|, |\eta|\} \leq M \in \mathbb{R}_+$, *there exists* $d = d_M(x) \in L^p(\mathcal{X})$ *such that*

$$|f(t, x, \xi) - f(s, x, \eta)| \leq d\left(|t - s|^{\gamma} + |\xi - \eta|^{\rho}\right),$$

for some $\gamma, \rho > 0$.

Then there exists a unique strong solution of (3.1.1) *in the sense of Definition 3.2.1 on the time interval* $(0, T_{\max})$, *such that either* $T_{\max} = \infty$, *or* $T_{\max} < \infty$ *and* (3.1.31) *is satisfied with* $p_0 = \infty$. *This strong solution also satisfies*

$$u \in C^{0,\kappa}([0, T_{\max}); D((-A_p)^{\theta})) \cap C((0, T_{\max}); D(A_p)). \tag{3.2.4}$$

for some real number $\kappa > 0$.

Proof For $p \in (\beta_A, \infty)$, let $\eta \in (\beta_A/p, 1)$ such that $\eta < \theta$, and a sufficiently small $\mu \in (0, 1)$, $\eta + \mu = \theta$. By Proposition 2.2.7, $u_0 \in D(A_p) \hookrightarrow D((-A_p)^{\theta}) \hookrightarrow L^{\infty}(\mathcal{X})$. Notice that (3.2.2) implies that we are in the assumptions of Theorem 3.1.4 since $\alpha - \mathfrak{n}/q_1 - 1/q_2 > \alpha\theta + \mathfrak{n}/p > 0$ (recall $\mathfrak{n} = \beta_A \alpha$). Hence, by application of Theorem 3.1.10 (or Corollary 3.1.11) there exists a (unique) mild solution $u \in E_{\infty,0,T}$, $T \in (0, T_{\max})$, that is given by an integral solution (see Definition 3.1.1-(d)) that also satisfies (3.1.32) on $[0, T_{\max})$. When (3.2.3) is in full force and instead $p \leq q_1$ we obtain the same conclusion. Next, in view of assumption (**HA**) for the operator A, we also recall from [17, p. 26] (since the semigroup S is analytic) that for all $t > 0$,

$$\left\|(-A_p)^{-(1-\tau)}(S(t) - I)\right\|_{p,p} \leq C_p t^{1-\tau}, \text{ for all } \tau \in (0, 1) \tag{3.2.5}$$

and

$$\left\|(-A_p)^{\tau} S(t)\right\|_{p,p} \leq C_p t^{-\tau}, \text{ for all } \tau \in [0, 1]. \tag{3.2.6}$$

In all the estimates that follow, we let $0 < t \leq t + h \leq T < T_{\max}$ such that $h \in [0, T - t]$ (w.l.o.g, we assume that $h \leq 1$). We observe preliminarily that since $\eta < \theta$ and $q_2 \in (1/\alpha, \infty]$, the first alternative (a) yields

$$\alpha(1 - \eta) - \mathfrak{n}\left(\frac{1}{q_1} - \frac{1}{p}\right) - \frac{1}{q_2} \overset{(3.2.2)}{>} 0 \tag{3.2.7}$$

[1] Here and everywhere else, by the statement "for almost all $x \in \mathcal{X}$" it is understood that $x \in \mathcal{X} \backslash \mathcal{J}$, where $\mathfrak{m}(\mathcal{J}) = 0$.

and

$$\frac{1}{q_2}+\frac{n}{q_1}<\alpha,\ \eta+\beta_A\left(\frac{1}{q_1}-\frac{1}{p}\right)<\theta+\beta_A\left(\frac{1}{q_1}-\frac{1}{p}\right)\overset{(3.2.2)}{<}1. \tag{3.2.8}$$

On the other hand, from assumption (b) of the theorem we infer that

$$\alpha\,(1-\eta)-\frac{1}{q_2}\overset{(3.2.3)}{>}0 \tag{3.2.9}$$

as well as the first of (3.2.8) holds once again.

Step 1 (Uniform estimates for S_α, P_α, $\alpha\in(0,1)$). Based on definition (2.1.9), we have

$$P_\alpha((2t)^{1/\alpha})=\alpha\,(2t)^{(\alpha-1)/\alpha}\int_0^\infty \tau\Phi_\alpha(\tau)S(2\tau t)d\tau,$$

for all $t>0$. The semigroup property $S\,(2\tau t)=S\,(t\tau)\,S\,(t\tau)$ and the ultracontractivity estimate for S (see (2.2.3)) imply that

$$\begin{aligned}
&(2t)^{\frac{1}{\alpha}-1}\left\|(-A_p)^\eta P_\alpha((2t)^{1/\alpha})\right\|_{p,q_1} \\
&\leq\int_0^\infty \tau\Phi_\alpha(\tau)\left\|(-A_p)^\eta S(\tau t)\right\|_{p,p}\|S\,(\tau t)\|_{p,q_1}d\tau \\
&\overset{(3.2.6)}{\leq} C\,(p,\eta)\,t^{-\eta-\beta_A\left(\frac{1}{q_1}-\frac{1}{p}\right)}\int_0^\infty \tau^{1-\eta-\beta_A\left(\frac{1}{q_1}-\frac{1}{p}\right)}\Phi_\alpha(\tau)d\tau \\
&\leq C\,(p,\eta,\beta_A,q_1)\,t^{-\eta-\beta_A\left(\frac{1}{q_1}-\frac{1}{p}\right)},
\end{aligned} \tag{3.2.10}$$

since the last integral in (3.2.10) is finite owing to the second of (3.2.8) (see once again (2.1.8)). The constant $C=C\,(p,\eta,\beta_A,q_1)$ is bounded as $\alpha\to 1$. Rescaling $t\mapsto t^\alpha/2$ in this estimate, we derive

$$t^{1-\alpha}\left\|(-A_p)^\eta P_\alpha(t)\right\|_{p,q_1}\leq Ct^{-\eta\alpha-\beta_A\alpha\left(\frac{1}{q_1}-\frac{1}{p}\right)},\ \text{whenever } p\geq q_1. \tag{3.2.11}$$

On the other hand, when $p\leq q_1$ we have $\|S\,(t)\|_{p,q_1}\leq\|S\,(t)\|_{p,p}\leq C$. Arguing in a similar fashion to estimate (3.2.10), the analogue of (3.2.11) then reads

$$t^{1-\alpha}\left\|(-A_p)^\eta P_\alpha\,(t)\right\|_{p,q_1}\leq Ct^{-\eta\alpha},\ \text{whenever } p\leq q_1. \tag{3.2.12}$$

Let us now recall $\theta \in (\beta_A/p, 1)$. The analytic estimates (3.2.5)–(3.2.6) also imply the following estimates for all $t > 0$,

$$\left\| (-A_p)^{-(1-\theta)} (S_\alpha(t) - I) \right\|_{p,p} \leq Ct^{\alpha(1-\theta)} \tag{3.2.13}$$

and

$$t^{1-\alpha} \left\| (-A_p)^{\theta} P_\alpha(t) \right\|_{p,p} \leq Ct^{-\alpha\theta}; \tag{3.2.14}$$

they follow easily owing to the definition (2.1.9) for the operators S_α, P_α, $\alpha \in (0, 1)$.

Step 2 (Regularity properties for the mild solution). Since every mild solution is an integral solution, we have

$$u(t+h) - u(t) = (S_\alpha(t+h) - S_\alpha(t)) u_0 + \int_t^{t+h} P_\alpha(t+h-\tau) f(\tau, u(\tau)) d\tau \tag{3.2.15}$$

$$+ \int_0^t (P_\alpha(t+h-\tau) - P_\alpha(t-\tau)) f(\tau, u(\tau)) d\tau.$$

We now check that the initial condition is satisfied at least in the sense of (3.2.1). By virtue of the bound $u \in L^\infty((0, T); L^\infty(\mathcal{X}))$ with $T < T_{\max}$ (indeed, $\sup_{t\in[0,T]} \|u\|_{L^\infty(\mathcal{X})} \leq U$, $u \in E_{\infty,0,T}$) and assumption (**F3**), we deduce for $t \in [0, 1]$,

$$\begin{aligned}
&\|u(t) - u_0\|_{L^\infty(\mathcal{X})} \\
\leq& C \left\| (-A_p)^{\theta} (u(t) - u_0) \right\|_{L^p(\mathcal{X})} \\
\leq& C \left\| (-A_p)^{\theta} (S_\alpha(t) u_0 - u_0) \right\|_{L^p(\mathcal{X})} \\
&+ C \int_0^t \left\| (-A_p)^{\theta} P_\alpha(t-\tau) \right\|_{p,q_1} \|f(\tau, u(\tau))\|_{L^{q_1}(\mathcal{X})} d\tau \\
\leq& C \left\| (-A_p)^{-(1-\theta)} (S_\alpha(t) - I) \right\|_{p,p} \|A_p u_0\|_{L^p(\mathcal{X})} \\
&+ C \int_0^t \left\| (-A_p)^{\theta} P_\alpha(t-\tau) \right\|_{p,q_1} \|c(\cdot, \tau)\|_{L^{q_1}(\mathcal{X})} d\tau Q(U) \\
\overset{(3.2.13)}{\leq}& Ct^{\alpha(1-\theta)} \|u_0\|_{D(A_p)} + Q(U) \left(\int_0^t \left\| (-A_p)^{\theta} P_\alpha(t-\tau) \right\|_{p,q_1}^{\overline{q}} d\tau \right)^{\frac{1}{\overline{q}}} C \|c\|_{q_1,q_2},
\end{aligned} \tag{3.2.16}$$

by Hölder's inequality where $\overline{q}(1 - 1/q_2) = 1$. The first summand on the right-hand side of (3.2.16) clearly tends to zero as $t \to 0^+$. For the second summand, we argue slightly differently according to the cases whether $p \geq q_1$ or $p \leq q_1$, respectively. Let us first assume $p \geq q_1$ and set $\chi := \alpha - 1 - \theta\alpha - \beta_A \alpha \left(\frac{1}{q_1} - \frac{1}{p}\right)$. By virtue of the estimate (3.2.11) we have then

$$\left(\int_0^t \left\|(-A_p)^\theta P_\alpha (t-\tau)\right\|_{p,q_1}^{\overline{q}} d\tau\right)^{\frac{1}{\overline{q}}} \leq C\left(\int_0^t (t-\tau)^{(\alpha-1)\overline{q} - \theta\alpha\overline{q} - \beta_A\alpha\left(\frac{1}{q_1}-\frac{1}{p}\right)\overline{q}} d\tau\right)^{\frac{1}{\overline{q}}}$$
$$= \frac{Ct^{\chi+1/\overline{q}}}{(\chi\overline{q}+1)^{1/\overline{q}}} \to 0, \tag{3.2.17}$$

as $t \to 0^+$, since $\chi + 1/\overline{q} > 0 \Leftrightarrow \chi - 1/q_2 > -1$, where the latter is also equivalent to assumption (3.2.2). In the other case $p \leq q_1$, we set $\chi := \alpha - 1 - \theta\alpha$ and exploit the estimate (3.2.12) instead. Namely, we get

$$\left(\int_0^t \left\|(-A_p)^\theta P_\alpha (t-\tau)\right\|_{p,q_1}^{\overline{q}} d\tau\right)^{1/\overline{q}} \leq C\left(\int_0^t (t-\tau)^{\chi\overline{q}} d\tau\right)^{\frac{1}{\overline{q}}} \tag{3.2.18}$$
$$= \frac{Ct^{\chi+1/\overline{q}}}{(\chi\overline{q}+1)^{\frac{1}{\overline{q}}}} \to 0, \text{ as } t \to 0^+,$$

since $\chi + 1/\overline{q} > 0 \Leftrightarrow \alpha(1-\theta) > 1/q_2$, which is satisfied by (3.2.3). We thus conclude that the mild solution satisfies (3.2.1) (as well as $\lim_{t\to 0^+} u(t) = u_0$ in the $D\left((-A_p)^\theta\right)$-norm) for every $u_0 \in D\left(A_p\right)$, with $p \in (\beta_A, \infty)$.

We claim next that $u \in C^{0,\kappa}((0,T); L^\infty(\mathcal{X}))$, for some $\kappa > 0$. We consider $\alpha \in (0,1)$ since the case $\alpha = 1$ follows with minor (and straight-forward) modifications. More precisely, in that case we can further take advantage of the fact that S is also a semigroup, as well as of the simple identity

$$u(t+h) - u(t) = (S(h) - I)\, S(t)\, u_0 + \int_0^h S(h-\tau)\, f(t+\tau, u(t+\tau))\, d\tau.$$

Let us recall that $\eta, \theta \in (\beta_A/p, 1)$ and $\eta < \theta = \eta + \mu$. In particular, it still holds $D\left((-A_p)^\eta\right) \hookrightarrow L^\infty(\mathcal{X})$. We now estimate the first summand in (3.2.15), based on the definition for the operator S_α,

$$\left\|\left(S_\alpha(t+h) - S_\alpha(t)\right)u_0\right\|_{L^\infty(\mathcal{X})} \tag{3.2.19}$$
$$\leq \left\|(-A_p)^\eta \left(S_\alpha(t+h) - S_\alpha(t)\right) u_0\right\|_{L^p(\mathcal{X})}$$
$$\leq \int_0^\infty \Phi_\alpha(\tau) \left\|(-A_p)^\eta \left(S\left(\tau(t+h)^\alpha\right) - S\left(\tau t^\alpha\right)\right)u_0\right\|_{L^p(\mathcal{X})} d\tau$$
$$\leq \int_0^\infty \Phi_\alpha(\tau) \left\|(-A_p)^{-\mu}\left(S\left(\tau(t+h)^\alpha - \tau t^\alpha\right) - I\right)\right\|_{p,p} \left\|(-A_p)^{\mu+\eta} S\left(\tau t^\alpha\right) u_0\right\|_{L^p(\mathcal{X})} d\tau$$

$$\overset{(3.2.5)}{\leq} C(p,\mu)\int_0^\infty \Phi_\alpha(\tau)\left(\tau(t+h)^\alpha-\tau t^\alpha\right)^\mu \left\|(-A_p)^\theta S\left(\tau t^\alpha\right)u_0\right\|_{L^p(\mathcal{X})}d\tau$$

$$\leq C(p,\mu,\theta)\left\|A_p u_0\right\|_{L^p(\mathcal{X})}\left(\int_0^\infty \Phi_\alpha(\tau)\tau^\mu d\tau\right)\left(C_T h^{\mu/s}\right)$$

$$\leq C(p,\mu,\theta,T)\left\|u_0\right\|_{D(A_p)}h^{\mu/s},$$

for some $1/s < \alpha$; here, we have also exploited the $D\left(A_p\right)$-contractivity of the operator S and the application of Lemma A.0.4-(i) with $\varepsilon := \alpha$. In the second inequality we have also used the semigroup property

$$\begin{aligned} S\left(\tau(t+h)^\alpha\right)-S\left(\tau t^\alpha\right) &= S\left(\tau(t+h)^\alpha-\tau t^\alpha\right)S\left(\tau t^\alpha\right)-S\left(\tau t^\alpha\right) \\ &= \left(S\left(\tau(t+h)^\alpha-\tau t^\alpha\right)-I\right)S\left(\tau t^\alpha\right). \end{aligned} \tag{3.2.20}$$

We deal now with the second summand in (3.2.15). Assume the first alternative (a) (when $p \geq q_1$) and recall (3.2.7)–(3.2.8) hold. By assumption (**F3**), we have

$$\left\|\int_t^{t+h} P_\alpha(t+h-\tau)f(\tau,u(\tau))d\tau\right\|_{D(A_p^\eta)} \tag{3.2.21}$$

$$\leq \int_t^{t+h}\left\|(-A_p)^\eta P_\alpha(t+h-\tau)\right\|_{p,q_1}\|f(\tau,u(\tau))\|_{L^{q_1}(\mathcal{X})}d\tau$$

$$\leq \int_t^{t+h}\left\|(-A_p)^\eta P_\alpha(t+h-\tau)\right\|_{p,q_1}\|c(\cdot,\tau)\|_{L^{q_1}(\mathcal{X})}d\tau\, Q(U)$$

$$\overset{(3.2.11)}{\leq} CQ(U)\left(\int_t^{t+h}(t+h-\tau)^{(\alpha(1-\eta)-1-\beta_A\alpha(1/q_1-1/p))\overline{q}}d\tau\right)^{\frac{1}{\overline{q}}}\|c\|_{q_1,q_2},$$

by Hölder's inequality where $\overline{q}\,(1-1/q_2)=1$. Notice that for

$$\xi := \alpha(1-\eta)-1-\beta_A\alpha\left(1/q_1-1/p\right),$$

we have that $\xi\overline{q}+1>0 \Leftrightarrow \xi-1/q_2>-1$, in light of (3.2.7). Therefore, since $D\left((-A_p)^\eta\right)\hookrightarrow L^\infty(\mathcal{X})$ from (3.2.21) we immediately deduce

$$\left\|\int_t^{t+h} P_\alpha(t+h-\tau)f(\tau,u(\tau))d\tau\right\|_{L^\infty(\mathcal{X})} \leq \left(Q(U)\|c\|_{q_1,q_2}\right)\frac{Ch^{\xi+1/\overline{q}}}{(\xi\overline{q}+1)^{1/\overline{q}}}. \tag{3.2.22}$$

When the second alternative (b) holds with $p \leq q_1$, we employ the estimate (3.2.12) instead of (3.2.11). Similarly to the derivation of (3.2.22), with a different value

$\xi := \alpha(1-\eta) - 1$ (such that $\xi\overline{q} + 1 > 0 \Leftrightarrow$ condition (3.2.9)), we arrive at the estimate

$$\begin{aligned}
&\left\| \int_t^{t+h} P_\alpha (t+h-\tau) f(\tau, u(\tau)) \, d\tau \right\|_{L^\infty(\mathcal{X})} &(3.2.23)\\
&\leq Q(U) \left(\int_t^{t+h} C (t+h-\tau)^{(\alpha(1-\eta)-1)\overline{q}} \, d\tau \right)^{1/\overline{q}} \|c\|_{q_1,q_2} \\
&\leq \left(Q(U) \|c\|_{q_1,q_2} \right) \frac{C h^{\xi + 1/\overline{q}}}{(\xi\overline{q}+1)^{1/\overline{q}}}.
\end{aligned}$$

Finally, we estimate the last and most difficult summand in (3.2.15). We begin with the following identity which holds in light of the definition (2.1.9) for the operator P_α:

$$\begin{aligned}
&P_\alpha (t+h-\tau) - P_\alpha (t-\tau) &(3.2.24)\\
=&\alpha (t+h-\tau)^{\alpha-1} \int_0^\infty \sigma \Phi_\alpha(\sigma) \left(S\left(\sigma (t+h-\tau)^\alpha\right) - S\left(\sigma (t-\tau)^\alpha\right)\right) d\sigma \\
&+ \alpha \left((t+h-\tau)^{\alpha-1} - (t-\tau)^{\alpha-1} \right) \int_0^\infty \sigma \Phi_\alpha(\sigma) S\left(\sigma(t-\tau)^\alpha\right) d\sigma \\
=:&L_h(t,\tau) + K_h(t,\tau).
\end{aligned}$$

Arguing in a similar fashion as we did in the estimate (3.2.19), taking advantage of the semigroup property (3.2.20) and the fact that $\left\| (-A_p)^\eta S(t) \right\|_{p,q_1} \leq C t^{-\eta - \beta_A(1/q_1 - 1/p)}$, we get

$$\begin{aligned}
&\left\| (-A_p)^\eta L_h(t,\tau) \right\|_{p,q_1} &(3.2.25)\\
&\leq C\alpha (t+h-\tau)^{\alpha-1} (t-\tau)^{-\alpha\eta - \beta_A \alpha(1/q_1 - 1/p)} \left(\int_0^\infty \sigma^{1+\mu-\theta-\beta_A(1/q_1-1/p)} \Phi_\alpha(\sigma) \, d\sigma \right) \\
&\times \left((t+h-\tau)^\alpha - (t-\tau)^\alpha \right)^\mu \\
&\overset{\text{(Lemma A.0.4-(i))}}{\leq} C_T (t+h-\tau)^{\alpha-1} (t-\tau)^{-\alpha\eta - \beta_A\alpha(1/q_1-1/p)} h^{\mu/\tau_0},
\end{aligned}$$

for some $1/\tau_0 < \alpha$, provided that we are in the assumptions of the first alternative (a). With the second alternative (b), the corresponding estimate reads

$$\left\| (-A_p)^\eta L_h(t,\tau) \right\|_{p,q_1} \leq C_T (t+h-\tau)^{\alpha-1} (t-\tau)^{-\alpha\eta} h^{\mu/\tau_0}. \tag{3.2.26}$$

Let us set $\varpi(\eta) := -\alpha\eta - \beta_A\alpha(1/q_1 - 1/p) < 0$ if $p \geq q_1$ and $\varpi(\eta) := -\alpha\eta < 0$ if $p \leq q_1$. By virtue of the uniform estimates (3.2.25)–(3.2.26) we find for $t \in (0, 1]$ (the proof of the case $t > 1$ can be reduced to the case $t \in (0, 1]$, by choosing $k \in \mathbb{N}$ such that $k < t \leq k + 1$ and by arguing exactly as in the proof of Lemma A.0.1),

$$\begin{aligned}
&\left\| \int_0^t L_h(t,\tau) f(\tau, u(\tau))\, d\tau \right\|_{D(A_p^\eta)} \\
&\leq \int_0^t \left\| (-A_p)^\eta L_h(t,\tau) \right\|_{p,q_1} \| f(\tau, u(\tau)) \|_{L^{q_1}(\mathcal{X})}\, d\tau \\
&\leq Q(U) \|c\|_{q_1,q_2} \left(\int_0^t \left\| (-A_p)^\eta L_h(t,\tau) \right\|_{p,q_1}^{\overline{q}} d\tau \right)^{\frac{1}{\overline{q}}} \\
&\leq C_T h^{\mu/\tau_0} \left(\int_0^t (t+h-\tau)^{\overline{q}(\alpha-1)} (t-\tau)^{\varpi\overline{q}}\, d\tau \right)^{\frac{1}{\overline{q}}} Q(U) \|c\|_{q_1,q_2},
\end{aligned}$$

where once again $\overline{q}(1 - 1/q_2) = 1$. Since $(t + h - \tau)^{(\alpha-1)\overline{q}} \leq (t - \tau)^{(\alpha-1)\overline{q}}$, for $0 \leq \tau < t \leq t + h \leq T$, the previous estimate then implies that

$$\begin{aligned}
&\left\| \int_0^t L_h(t,\tau) f(\tau, u(\tau))\, d\tau \right\|_{D((-A_p)^\eta)} \qquad (3.2.27)\\
&\leq C_T h^{\mu/\tau_0} Q(U) \|c\|_{q_1,q_2} \left(\int_0^t (t-\tau)^{(\varpi+\alpha-1)\overline{q}}\, d\tau \right)^{\frac{1}{\overline{q}}} \\
&\leq C_T(\omega, \beta_A, \alpha, q_1, p)\, h^{\mu/\tau_0} Q(U) \|c\|_{q_1,q_2} T^{\alpha+\varpi-\frac{1}{q_2}}.
\end{aligned}$$

We observe that $\alpha + \varpi(\eta) - 1/q_2 > 0$ in both cases of (a)–(b) due to (3.2.7)–(3.2.9). Concerning the second summand in (3.2.24) with the same value $\varpi = \varpi(\eta) < 0$ as above, we find

$$\begin{aligned}
&\left\| (-A_p)^\eta K_h(t,\tau) \right\|_{p,q_1} \qquad (3.2.28)\\
&\leq \alpha \left| (t+h-\tau)^{\alpha-1} - (t-\tau)^{\alpha-1} \right| \int_0^\infty \sigma \Phi_\alpha(\sigma) \left\| (-A_p)^\eta S\left(\sigma (t-\tau)^\alpha\right) \right\|_{p,q_1} d\sigma \\
&\leq C(t-\tau)^\varpi \left(\int_0^\infty \sigma^{1-\varpi/\alpha} \Phi_\alpha(\sigma)\, d\sigma \right) \left((t-\tau)^{\alpha-1} - (t+h-\tau)^{\alpha-1} \right) \\
&\leq C(\varpi, \alpha)(t-\tau)^\varpi \left((t-\tau)^{\alpha-1} - (t+h-\tau)^{\alpha-1} \right).
\end{aligned}$$

In view of this estimate and the Hölder inequality with $\overline{q}(1-1/q_2) = 1$, $r_2(1-1/r_1) = 1$, we obtain

$$\left\| \int_0^t K_h(t,\tau) f(\tau,u(\tau))\,d\tau \right\|_{D((-A_p)^\eta)} \tag{3.2.29}$$

$$\leq Q(U)\|c\|_{q_1,q_2} \left(\int_0^t C(t-\tau)^{\varpi\overline{q}} \left((t-\tau)^{\alpha-1} - (t+h-\tau)^{\alpha-1} \right)^{\overline{q}} d\tau \right)^{\frac{1}{\overline{q}}}$$

$$\leq Q(U)\|c\|_{q_1,q_2} \left(\int_0^t C(t-\tau)^{\varpi\overline{q}r_1}\,d\tau \right)^{\frac{1}{\overline{q}r_1}} \times$$

$$\times \left(\int_0^t \left((t-\tau)^{\alpha-1} - (t+h-\tau)^{\alpha-1} \right)^{\overline{q}r_2} d\tau \right)^{\frac{1}{\overline{q}r_2}}$$

$$\leq Q(U)\|c\|_{q_1,q_2} \left(\int_0^t C(t-\tau)^{\varpi\overline{q}r_1}\,d\tau \right)^{\frac{1}{\overline{q}r_1}} \times$$

$$\times \left(\int_0^t \left((t-\tau)^{\alpha-1} - (t+h-\tau)^{\alpha-1} \right)^{\overline{q}r_2} d\tau \right)^{\frac{1}{\overline{q}r_2}}.$$

Notice that by (3.2.7)–(3.2.9), there exists a sufficiently small $\varepsilon \in (0, \alpha - 1/q_2)$ such that $\alpha + \varpi - 1/q_2 > \varepsilon > 0$. To this end, select $r_1 < \infty$ such that $1/(\overline{q}r_1) = \alpha - 1/q_2 - \varepsilon > 0$; clearly $1/(\overline{q}r_2) = 1 - \alpha + \varepsilon > 0$ and $(\alpha-1)\overline{q}r_2 + 1 = \varepsilon/(1-\alpha+\varepsilon) > 0$. By application of Lemma A.0.4-(ii), we get for $t \leq T$,

$$\left(\int_0^t C(t-\tau)^{\varpi\overline{q}r_1}\,d\tau \right)^{\frac{1}{\overline{q}r_1}} \left(\int_0^t \left((t-\tau)^{\alpha-1} - (t+h-\tau)^{\alpha-1} \right)^{\overline{q}r_2} d\tau \right)^{\frac{1}{\overline{q}r_2}} \tag{3.2.30}$$

$$\leq C(\overline{q}, r_1, r_2, \varpi)\, T^{\varpi+1/\overline{q}r_1} \left(\int_0^t (t-\tau)^{(\alpha-1)\overline{q}r_2} - (t+h-\tau)^{(\alpha-1)\overline{q}r_2}\,d\tau \right)^{\frac{1}{\overline{q}r_2}}$$

$$\leq C(\overline{q}, r_1, r_2, \varpi)\, T^{\varpi+1/\overline{q}r_1} \left(-t^{\alpha-1+1/\overline{q}r_2} - h^{\alpha-1+1/\overline{q}r_2} + (t+h)^{\alpha-1+1/\overline{q}r_2} \right)$$

$$\leq C(\overline{q}, r_1, r_2, \varpi)\, T^{\varpi+1/\overline{q}r_1} \left(h^{\alpha-1+1/\overline{q}r_2} + h^{1/\overline{s}} \right),$$

for some $1/\overline{s} < \alpha - 1 + 1/\overline{q}r_2 = \varepsilon$, owing also to the application of Lemma A.0.4-(i). Inserting (3.2.30) into the estimate (3.2.29), we infer that

$$\left\| \int_0^t K_h(t,\tau) f(\tau,u(\tau))\,d\tau \right\|_{D((-A_p)^\eta)}$$

$$\leq C T^{\varpi+1/\overline{q}r_1} \left(h^{\alpha-1+1/\overline{q}r_2} + h^{1/\overline{s}} \right) Q(U)\|c\|_{q_1,q_2}. \tag{3.2.31}$$

Collecting all the uniform estimates (3.2.19), (3.2.27) and (3.2.31) and recalling the identity (3.2.15), we obtain

$$\|u(t+h) - u(t)\|_{L^\infty(\mathcal{X})} \leq C_T Q(U) \|c\|_{q_1,q_2} h^\kappa,$$

for $\kappa := \min(\mu/s, \xi + 1/\overline{q}, 1/\overline{s}) > 0$, which gives the desired regularity

$$u \in C^{0,\kappa}\left([0,T]; L^\infty(\mathcal{X})\right). \tag{3.2.32}$$

Step 3 (Final argument). We are now ready to conclude the proof. For any mild solution of Problem (3.1.1), we define $H(\cdot,t) : [0,T] \to L^p(\mathcal{X})$, as $H(\cdot,t) := f(t,\cdot,u(\cdot,t))$ where we recall that the locally Lipschitz function f obeys the assumptions (**F3**)–(**F4**). By the additional assumption (**F5**) of the theorem, we then have $H(\cdot,t) \in C^{0,\sigma}((0,T); L^p(\mathcal{X}))$, for some $\sigma > 0$, by means of (3.2.32). Note that the (mild) integral solution can be also written as

$$u(\cdot,t) = S_\alpha(t) u_0 + \int_0^t P_\alpha(t-\tau) H(\cdot,\tau) d\tau, \ \alpha \in (0,1)$$

and

$$u(\cdot,t) = S(t) u_0 + \int_0^t S(t-\tau) H(\cdot,\tau) d\tau, \ \alpha = 1,$$

for all $t \in [0,T]$, $T < T_{\max}$. Hence, in view of this formula and the application of Theorem 2.1.7 with the choice $Y = L^p(\mathcal{X})$, we can infer the remaining properties (c), (d) of Definition 3.2.1. We have verified that u is a strong solution in the sense of Definition 3.2.1. The proof of the theorem is finished. □

The additional assumption (**F5**) of Theorem 3.2.2 can be essentially dropped in some special (albeit interesting) cases provided that $f(t,x,\xi)$ is independent of t.

Corollary 3.2.3 *Let $u_0 \in D(A_p)$ for some $p \in (\beta_A, \infty) \cap (1,\infty)$ and assume that $f = f(x,\xi)$ satisfies conditions* (**F3**)–(**F4**) *for some function $c = c(x) \in L^p(\mathcal{X})$. Then there exists a unique strong solution to Problem* (3.1.1) *in the sense of Definition 3.2.1 on the time interval* $(0, T_{\max})$, *such that either $T_{\max} = \infty$, or $T_{\max} < \infty$ and* (3.1.31) *is satisfied with $p_0 = \infty$. The strong solution also satisfies* (3.2.4).

Proof Indeed, observe that $c = c(x) \in L_{p,\infty}$, with $q_1 := p$, $q_2 := \infty$, and the second alternative (b) of Theorem 3.2.2 is automatically satisfied. By assumption (**F4**) together with the regularity property (3.2.32) we can then infer once again that $H(\cdot,t) = f(\cdot,u(\cdot,t)) \in C^{0,\sigma}((0,T); L^p(\mathcal{X}))$, for some $\sigma > 0$. By following the same argument from Step 3 in the proof of Theorem 3.2.2, we easily arrive at the desired conclusion. □

We view the regularity property (b) of Definition 3.2.1 as "minimally smooth", that our strong solution possesses under the general assumption (**HA**). When more detailed information is a priori known on the operator A, such smoothness can be essentially improved beyond the $L^\infty(\mathcal{X})$-spatial regularity.

Remark 3.2.4 Indeed, let $\mathcal{X} = \Omega \subset \mathbb{R}^N$ be an open set having a boundary of class C^2, and A is a sectorial operator in $Y = L^p(\Omega)$, $p \in (1, \infty)$ with $D(A_p) \subset W^{2,p}(\Omega)$ and $\beta_A = N/2$. In particular, all the operators defined previously in Example 2.3.5 (see (a)–(d)) satisfy these assumptions. By Proposition 2.2.7, there exists a small $\nu > 0$ such that $\theta > \nu/2 + N/(2p)$. It follows from [17, Theorem 1.6.1] that $D((-A_p)^\theta) \hookrightarrow C^{0,\nu}(\overline{\Omega})$ and then by virtue of Theorem 3.2.2, that every strong solution also satisfies

$$u \in C^{0,\kappa}\left([0,T]; C^{0,\nu}\left(\overline{\Omega}\right)\right), \text{ for some } \kappa, \nu > 0 \tag{3.2.33}$$

and

$$u \in C\left((0,T]; W^{2,p}(\Omega)\right), \tag{3.2.34}$$

where $T < T_{\max}$.

Remark 3.2.5 We also mention the following.

(*a*) For the operators A defined in Example 2.3.8 (a)–(c) we have the following situation. Recall that $A = A_2 = -(-B)^s$ where B is as in Remark 3.2.4 above and $0 < s < 1$. In that case (always under the assumption that Ω is smooth and the coefficients of the initial operator are also smooth) we have that $D(A_p) \subset W^{2s,p}(\Omega)$ for every $p \in (1, \infty)$ (see e.g. [23, Theorem 7.1] or [16]) and $\beta_A = \frac{N}{2s}$. Let $\theta > 0$. Since $(-A)^\theta = (-B)^{s\theta}$, it follows from Proposition 2.2.7 that there exists a small $\nu_s > 0$ such that $\theta > \nu_s/2 + N/(2sp)$. Hence, by [17, Theorem 1.6.1] we have that $D((-A_p)^\theta) \hookrightarrow C^{0,\nu_s}(\overline{\Omega})$ and then by virtue of Theorem 3.2.2, it follows that every strong solution also satisfies

$$u \in C^{0,\kappa_s}\left([0,T]; C^{0,\nu_s}\left(\overline{\Omega}\right)\right), \text{ for some } \kappa_s, \nu_s > 0 \tag{3.2.35}$$

and

$$u \in C\left((0,T]; W^{2s,p}(\Omega)\right), \tag{3.2.36}$$

where $T < T_{\max}$.

(*b*) For the operators given in Example 2.3.6, even if assuming that Ω is smooth, there is no global regularity results as the ones in part (a) available. In fact it is even known that in that case $D(A_p) \not\subset W^{2s,p}(\Omega)$. But most recently a local regularity result has been obtained in [2, 3] where the authors have shown that if $p \in [2, \infty)$ then $D(A_p) \subset W^{2s,p}_{\mathrm{loc}}(\Omega)$ and if $1 < p < 2$, then $D(A_p) \subset (B^s_{p,2})_{\mathrm{loc}}(\Omega)$. Therefore in those cases we will have that

$$\begin{cases} u \in C\left((0,T]; W^{2s,p}_{\mathrm{loc}}(\Omega)\right) & \text{if } 2 \leq p < \infty, \\ u \in C\left((0,T]; (B^s_{p,2})_{\mathrm{loc}}(\Omega)\right) & \text{if } 1 < p \leq 2, \end{cases}$$

where $T < T_{\max}$. Here $B^s_{p,2}(\mathbb{R}^N)$ denotes the Besov space.

We shall not pursue the issue of *spatial regularity* any further under the general assumption (**HA**). Nevertheless, we wanted to emphasize that such an additional regularity for the strong solution, like in (3.2.33)–(3.2.36), can be expected in more specific situations as a consequence of (3.2.4).

Finally, we conclude that every (maximal) bounded mild solution constructed by Theorem 3.1.10 (or any quasi-mild solution, as given by Corollary 3.1.11) regularizes to a strong solution for all positive times provided that q_1, q_2 and f satisfy the assumptions of Theorem 3.2.2.

Theorem 3.2.6 (Global Regularity of the Bounded Mild Solution) *Let u be the corresponding mild solution in the sense of Definition 3.1.1 (or Definition 3.1.3) on the interval $I = [0, T]$ or $I = [0, \infty)$ and let $M := \sup_{t \in I} \|u(\cdot, t)\|_{L^\infty(\mathcal{X})} < \infty$. Consider either alternative (a) or (b) of Theorem 3.2.2, along with assumption (**F5**) for the nonlinearity $f = f(t, x, \xi)$ (when f is independent of t, assume instead that $f = f(x, \xi)$ satisfies conditions (**F3**)–(**F4**) for some function $c = c(x) \in L^p(\mathcal{X})$, $p \in (\beta_A, \infty) \cap (1, \infty)$). Then for all $T_0 \in I \setminus \{0\}$, u is a strong solution on the time interval $I_0 := [T_0, T]$* (or $I_0 =: [T_0, \infty)$) *in the sense of Definition 3.2.1 (namely, (b)–(d) are satisfied for* any $p \in (\beta_A, \infty) \cap (1, \infty)$*).*

Proof For $\theta \in (\beta_A/p, 1)$, with $p \in (\beta_A, \infty) \cap (1, \infty)$, the formula for the integral solution allows to get the estimate

$$\begin{aligned} &\left\|(-A_p)^\theta u(\cdot, t)\right\|_{L^p(\mathcal{X})} \qquad (3.2.37)\\ &\leq \left\|(-A_p)^\theta S_\alpha(t) u_0\right\|_{L^p(\mathcal{X})} + Q(M) \int_0^t \left\|(-A_p)^\theta P(t-\tau)\right\|_{p,q_1} \|c(\cdot,\tau)\|_{L^{q_1}(\mathcal{X})} d\tau \\ &\leq \left\|(-A_p)^\theta S_\alpha(t)\right\|_{p,p} \|u_0\|_{L^p(\mathcal{X})} \\ &\quad + Q(M) \left(\int_0^t \left\|(-A_p)^\theta P(t-\tau)\right\|^{\overline{q}}_{p,q_1} d\tau\right)^{1/\overline{q}} \|c\|_{q_1,q_2}, \end{aligned}$$

where $\overline{q}(1 - 1/q_2) = 1$, for all $t \in [T_0, T]$. Exploiting the estimate (3.2.6) and recalling (3.2.17)–(3.2.18), we infer from (3.2.37) that

$$\left\|(-A_p)^\theta u(\cdot, t)\right\|_{L^p(\mathcal{X})} \leq C T_0^{-\theta\alpha} \|u_0\|_{L^p(\mathcal{X})} + Q(M) \|c\|_{q_1,q_2} \left(C T^{\chi + 1/\overline{q}}\right), \tag{3.2.38}$$

where $\chi + 1/\overline{q} > 0$, for all $T_0 \leq t \leq T$. Consider next the integral formula (3.2.15), which holds for all $T_0 \leq t < t+h \leq T$. Let $\eta < \theta = \eta + \mu$ such that $\eta \in (\beta_A/p, 1)$ and argue verbatim as in the proof of Theorem 3.2.2 (see Step 2, estimates (3.2.19)–(3.2.31)) with the exception of the first summand

$$\begin{aligned}
&\left\| (-A_p)^\eta \left(S_\alpha (t+h) - S_\alpha (t)\right) u_0 \right\|_{L^p(\mathcal{X})} \\
&\overset{(3.2.5)}{\leq} C(p,\mu) \int_0^\infty \Phi_\alpha(\tau) \left(\tau (t+h)^\alpha - \tau t^\alpha\right)^\mu \left\| (-A_p)^\theta S\left(\tau t^\alpha\right) u_0 \right\|_{L^p(\mathcal{X})} d\tau \\
&\leq C(p,\mu,\eta)\, t^{-\eta\alpha} \left(\int_0^\infty \Phi_\alpha(\tau) \tau^{-\eta} d\tau \right) \left(C_T h^{\mu/s}\right) \\
&\leq C(p,\mu,\theta,T)\, T_0^{-\eta\alpha} h^{\mu/s},
\end{aligned}$$

where $1/s < \alpha$. We once again derive

$$\begin{aligned}
&\left\| (-A_p)^\eta \left(u(\cdot, t+h) - u(\cdot, t)\right) \right\|_{L^p(\mathcal{X})} \qquad (3.2.39) \\
&\leq \left\| (-A_p)^\eta \left(S_\alpha (t+h) - S_\alpha (t)\right) u_0 \right\|_{L^p(\mathcal{X})} \\
&\quad + \int_t^{t+h} \left\| (-A_p)^\eta P_\alpha (t+h-\tau) \right\|_{p,q_1} \| f(\tau, u(\tau)) \|_{L^{q_1}(\mathcal{X})} d\tau \\
&\quad + \int_0^t \left\| (-A_p)^\eta \left(P_\alpha (t+h-\tau) - P_\alpha (t-\tau)\right) \right\|_{p,q_1} \| f(\tau, u(\tau)) \|_{L^{q_1}(\mathcal{X})} d\tau \\
&\leq \left(Q(M) \|c\|_{q_1,q_2} + C_T T_0^{-\eta\alpha} \right) h^\kappa,
\end{aligned}$$

with the same value $\kappa > 0$, see (3.2.32). The embedding $D((-A_p)^\eta) \hookrightarrow L^\infty(\mathcal{X})$ yields from estimate (3.2.39) the desired claim that u is κ-Hölder continuous on $[T_0, T]$ with respect to the $L^\infty(\mathcal{X})$-norm. Thus we may conclude the thesis exploiting the same step employed at the end of the proof of Theorem 3.2.2, on any time interval $[T_0, T] \subset I$. The proof is finished. □

3.3 Differentiability Properties in the Case $0 < \alpha < 1$

The problem of determining some additional smoothness for the strong solutions of the semilinear problem (3.1.1) is rather a complex one. Indeed in the case $\alpha \in (0,1)$, by Definition 3.2.1 and Theorem 3.2.2 each (maximally-defined) strong solution has the property

$$\partial_t^\alpha u \in C\left((0,T]; L^p(\mathcal{X})\right), \; u \in C^{0,\kappa}\left([0,T]; L^\infty(\mathcal{X})\right), \text{ for some } 0 < \kappa < 1, \tag{3.3.1}$$

for any $T < T_{\max}$. By Remark 2.1.5, the first of (3.3.1) implies that $u \in C^{0,\alpha}((0,T];L^p(\mathcal{X}))$ but u is *not* generally known to be in $C^1((0,T];L^p(\mathcal{X}))$. This is automatically true in the standard case when $\alpha = 1$. The existence of singularities in the derivative $u'(t)$ at $t = 0$ is immediately apparent from the notion of integral solution in (3.1.2); such singularities are present through the operator $P_\alpha(t)$ which now becomes unbounded as $t \to 0^+$ (in the case $\alpha = 1$, $P_\alpha(t) = S(t) = e^{At}$ is no longer singular as $t \to 0^+$). In the linear case (2.1.6), when the source $f = f(x,t)$ is independent of the variable u, the C^1-in time regularity can be found in the result of Proposition 2.1.9.

Our aim is to address the issue of C^1-regularity for the full semilinear problem (3.1.1) provided that $f = f(x,t,u)$ is smooth enough as a function in (t,u). We consider again the integral solution (3.1.2) and first compute its formal derivative

$$u'(t) = S_\alpha'(t)u_0 + P_\alpha(t)f(0,u(0)) + \int_0^t P_\alpha(t-\tau)\left[\partial_t f(\tau,u(\tau)) + \partial_u f(\tau,u(\tau))u'(\tau)\right]d\tau. \tag{3.3.2}$$

We also recall that $S_\alpha'(t)u_0 = P_\alpha(t)\left(A_p u_0\right)$ for $u_0 \in D\left(A_p\right)$. We can rewrite Eq. (3.3.2) in the form

$$V(t) = v(t) + \int_0^t P_\alpha(t-\tau)\partial_u f(\tau,u(\tau))V(\tau)d\tau \tag{3.3.3}$$

where we have set $V(t) = u'(t)$ and

$$v(t) := S_\alpha'(t)u_0 + P_\alpha(t)f(0,u(0)) + \int_0^t P_\alpha(t-\tau)\partial_t f(\tau,u(\tau))d\tau. \tag{3.3.4}$$

In the sequel, we will need some additional hypotheses on the nonlinear function f. But first, a preliminary lemma is required.

Lemma 3.3.1 *Let $p \in (1,\infty)$ and $T > 0$ be fixed but otherwise arbitrary. Consider the following pair of nonlinear equations*

$$w_j(t) = F_j(t) + \int_0^t P_\alpha(t-s)f_j\left(s,w_j(s)\right)ds,$$

for some $F_j \in L^1((0,T);L^p(\mathcal{X}))$ and assume that $w_j \in L^1((0,T);L^p(\mathcal{X}))$ exist a.e. in $(0,T)\times\mathcal{X}$, for $j = 1,2$. In addition assume the following:

(i) For almost all $x \in \mathcal{X}$, the functions $f_1(t,\xi)(x) = f_1(x,t,\xi)$ and $f_2(t,\xi)(x) = f_2(x,t,\xi)$ are continuous in $(t,\xi) \in [0,T]\times\mathbb{R}$.

(ii) For almost all $x \in \mathcal{X}$ and all $t \in [0,T]$, $f_1(x,t,\xi)$ is Lipschitz continuous in ξ with Lipschitz constant $L > 0$, independent of t and ξ.

Define

$$Q(t) := F_1(t) - F_2(t) + \int_0^t P_\alpha(t-\tau)\Big(f_1(\tau, w_2(\tau)) - f_2(\tau, w_2(\tau))\Big)d\tau.$$

Then for all $t \in (0, T]$ one has the estimate

$$\|w_1(t) - w_2(t)\|_{L^p(\mathcal{X})} \tag{3.3.5}$$

$$\leq \|Q(t)\|_{L^p(\mathcal{X})} + C\int_0^t (t-\tau)^{\alpha-1} E_{\alpha,\alpha}\left(c(t-\tau)^\alpha\right)\|Q(\tau)\|_{L^p(\mathcal{X})}\,d\tau,$$

for some $C, c > 0$ independent of t and w_j.

Proof Define $z := w_1 - w_2$, $F := F_1 - F_2$ and the function

$$G(t) := \begin{cases} \dfrac{f_1(t, w_1(t)) - f_1(t, w_2(t))}{z(t)}, & \text{if } z(t) \neq 0, \\ 0, & \text{if } z(t) = 0. \end{cases}$$

Then $G \in L^\infty((0, T); L^\infty(\mathcal{X}))$ and $\|G(t)\|_{L^\infty(\mathcal{X})} \leq L$, a.e. on $(0, T)$, by assumption (ii). Based on the definition of w_j, we have

$$z(t) = F(t) + \int_0^t P_\alpha(t-\tau)(f_1(\tau, w_2(\tau)) - f_2(\tau, w_2(\tau)))\,d\tau$$

$$+ \int_0^t P_\alpha(t-\tau)(f_1(\tau, w_1(\tau)) - f_1(\tau, w_2(\tau)))\,d\tau$$

$$= Q(t) + \int_0^t P_\alpha(t-\tau)\,G(\tau)\,z(s)\,d\tau$$

which then yields

$$\|z(t)\|_{L^p(\mathcal{X})} \leq \|Q(t)\|_{L^p(\mathcal{X})} + \int_0^t \|P_\alpha(t-\tau)\|_{p,p}\|G(\tau)\|_{L^\infty(\mathcal{X})}\|z(\tau)\|_{L^p(\mathcal{X})}\,d\tau \tag{3.3.6}$$

$$\leq \|Q(t)\|_{L^p(\mathcal{X})} + (C_p \cdot L)\int_0^t (t-\tau)^{\alpha-1}\|z(\tau)\|_{L^p(\mathcal{X})}\,d\tau,$$

owing to the bound $\|P_\alpha(t)\|_{p,p} \leq C_p t^{\alpha-1}$, for all $t > 0$. The final estimate (3.3.5) follows as an application of the (Gronwall) Lemma A.0.9 and (3.3.6). This concludes the proof of the lemma. □

We have the following result for each strong solution on $[0, T]$ (with $T < T_{\max}$ and $p \in (\beta_A, \infty) \cap (1, \infty)$) that is given by (3.1.2).

Theorem 3.3.2 *Let the assumptions of Theorem 3.2.2 be satisfied and assume the following hypotheses:*

(a) For almost all $x \in \mathcal{X}$ and all $t \in [0, T]$, the function $f(t, \xi)(x) = f(x, t, \xi)$ is continuously differentiable in $(t, \xi) \in [0, T] \times [-M, M]$, for some $M > 0$; moreover, for a.e. $x \in \mathcal{X}$ and $(t_j, \xi) \in [0, T] \times [-M, M]$ $(j = 1, 2)$, the estimate holds

$$|\partial_t f(x, t_1, \xi) - \partial_t f(x, t_2, \xi)| \leq d_M(x) |t_1 - t_2|^\beta, \tag{3.3.7}$$

for some $\beta > 0$ and $d_M \in L^p(\mathcal{X})$.

(b) For almost all $x \in \mathcal{X}$ and $t \in [0, T]$, the function $\partial_\xi f(t, \xi)(x) = \partial_\xi f(x, t, \xi)$ satisfies for all $\xi_1, \xi_2 \in [-M, M]$,

$$\left|\partial_\xi f(x, t, \xi_1) - \partial_\xi f(x, t, \xi_2)\right| \leq e_{M,T}(x) |\xi_1 - \xi_2|, \tag{3.3.8}$$

for some $e_{M,T} \in L^p(\mathcal{X})$.

Let V be a solution of (3.3.3) with v given by (3.3.4). Then the strong solution of Theorem 3.2.2 belongs to $C^1((0, T]; L^p(\mathcal{X}))$ and $u'(t) = V(t)$ on the interval $0 < t \leq T$.

Proof Let $\delta \in (0, T/3)$ be an arbitrarily small number and consider the right-difference

$$Z(t, h) := h^{-1}(u(t + h) - u(t)), \text{ for } h \in (0, \delta] \text{ and } 0 < t \leq T - \delta.$$

Notice that $Z(t - h, h)$ coincides with the left-difference. Moreover, since each strong solution is bounded, namely, $u \in L^\infty((0, T); L^\infty(\mathcal{X}))$, without loss of generality we may set

$$M := \|u\|_{L^\infty((0,T);L^\infty(\mathcal{X}))} < \infty.$$

It follows that $f(t, \xi)(x)$, $\partial_t f(t, \xi)(x)$ and $\partial_u f(t, \xi)(x)$ are all bounded (continuous) functions on $[0, T] \times [-M, M]$, as functions in $L^p(\mathcal{X})$, as well as $\partial_u f(t, \xi)(x)$ is globally Lipschitz continuous in $\xi \in [-M, M]$ on account of (3.3.8). In particular, let $K > 0$ be a bound for the following quantities

$$\sup_{t\in[0,T]} \|\partial_u f(x, t, u)\|_{L^p(\mathcal{X})} < \infty,$$

$$\sup_{t\in[0,T]} \left(\|\partial_t f(x, t, u)\|_{L^p(\mathcal{X})} + \|f(x, t, u)\|_{L^p(\mathcal{X})}\right) < \infty.$$

We recall once again that every strong solution u has the regularity (3.3.1). Since $u(t)$ also satisfies the integral equation (3.1.2), $Z(t,h)$ satisfies an equation of the form

$$Z(t,h) = R(t,h) + \int_0^t P_\alpha(t-\tau)\, \partial_u f\left(\tau, u^*(\tau)\right) Z(\tau,h)\, d\tau, \tag{3.3.9}$$

where $u^*(t)$ is between $u(t)$ and $u(t+h)$ and $\theta(h) \in (0,h)$, where

$$\begin{aligned} R(t,h) &:= h^{-1}\left(S_\alpha(t+h) - S_\alpha(t)\right) u_0 \\ &\quad + h^{-1} \int_t^{t+h} P_\alpha(\tau)\, f\left(t+h-\tau, u(t+h-\tau)\right) d\tau \\ &\quad + \int_0^t P_\alpha(\tau)\, \partial_t f\left(t+\theta(h)-\tau, u(t-\tau)\right) d\tau. \end{aligned}$$

We now apply the Gronwall Lemma 3.3.1 to the difference $Z(t,h) - V(t)$. For every $0 < t \le T-\delta$, we find

$$\begin{aligned} \|Z(t,h) - V(t)\|_{L^p(\mathcal{X})} &\le \|Q(t,h)\|_{L^p(\mathcal{X})} \\ &\quad + C\int_0^t (t-\tau)^{\alpha-1} E_{\alpha,\alpha}\left(c(t-\tau)^\alpha\right) \|Q(\tau,h)\|_{L^p(\mathcal{X})}\, d\tau, \end{aligned} \tag{3.3.10}$$

where

$$Q(t,h) := R(t,h) - v(t) + \int_0^t P_\alpha(t-\tau)\left(\partial_u f\left(\tau, u^*(\tau)\right) - \partial_u f\left(\tau, u(\tau)\right)\right) d\tau \tag{3.3.11}$$

The foregoing equation can be further split up into four distinct terms, as follows:

$$I(t,h) := \frac{S_\alpha(t+h) - S_\alpha(t)}{h} u_0 - S_\alpha'(t)\, u_0, \tag{3.3.12}$$

$$II(t,h) := h^{-1}\int_t^{t+h} P_\alpha(\tau)\, f\left(t+h-\tau, u(t+h-\tau)\right) d\tau - P_\alpha(t)\, f(0, u(0)), \tag{3.3.13}$$

$$III(t,h) := \int_0^t P_\alpha(\tau)\left(\partial_t f\left(t+\theta(h)-\tau, u(t-\tau)\right) - \partial_t f\left(t-\tau, u(t-\tau)\right)\right) d\tau, \tag{3.3.14}$$

$$IV(t,h) := \int_0^t P_\alpha(t-\tau)\left(\partial_u f\left(\tau, u^*(\tau)\right) - \partial_u f\left(\tau, u(\tau)\right)\right) d\tau. \tag{3.3.15}$$

To prove the claim of the theorem, we need to check first the following two steps:

(i) $\|Q(t,h)\|_{L^p(\mathcal{X})} \to 0$ in the limit of $h \to 0^+$, uniformly for $t \in [\delta, T-\delta]$.

(ii) $\|Q(\tau,h)\|_{L^p(\mathcal{X})} \leq K_0 + K_1\tau^{\alpha-1}$, for all $0 < \tau < t$, for some constants $K_0, K_1 > 0$ independent of t, h.

For (i), we have $\|I(t,h)\|_{L^p(\mathcal{X})} \to 0$ as $h \to 0^+$ uniformly in $t \in [\delta, T-\delta]$ since $S_\alpha(t)$ is analytic for $t \geq \delta > 0$ and $S_\alpha^{'}(t)u_0 = P_\alpha(t)(A_p u_0)$, $u_0 \in D(A_p)$. We use assumption (a) via the inequality (3.3.7), to estimate the third term:

$$\begin{aligned}
&\|III(t,h)\|_{L^p(\mathcal{X})} \qquad (3.3.16)\\
&\leq \int_0^t \|P_\alpha(\tau)\|_{p,p} \|\partial_t f(t+\theta(h)-\tau, u(t-\tau)) - \partial_t f(t-\tau, u(t-\tau))\|_{L^p(\mathcal{X})} d\tau\\
&\leq C_p \|d_M\|_{L^p(\mathcal{X})} \theta(h)^\beta \int_0^t \tau^{\alpha-1} d\tau\\
&\leq CT^\alpha \theta(h)^\beta,
\end{aligned}$$

which also goes to zero as $h \to 0^+$ (and so $\theta(h) \to 0$), uniformly for $t \in [\delta, T-\delta]$. We estimate the fourth term, owing to assumption (b) and (3.3.1),

$$\begin{aligned}
&\|IV(t,h)\|_{L^p(\mathcal{X})} \qquad (3.3.17)\\
&\leq \int_0^t \|P_\alpha(t-\tau)\|_{p,p} \left\|\partial_u f(\tau, u^*(\tau)) - \partial_u f(\tau, u(\tau))\right\|_{L^p(\mathcal{X})} d\tau\\
&\leq C_p \int_0^t (t-\tau)^{\alpha-1} \left\|e_{M,T}\right\|_{L^p(\mathcal{X})} \left\|u^*(\tau) - u(\tau)\right\|_{L^\infty(X)} d\tau\\
&\leq C_p \left\|e_{M,T}\right\|_{L^p(\mathcal{X})} \int_0^t (t-\tau)^{\alpha-1} \|u(\tau+h) - u(\tau)\|_{L^\infty(\mathcal{X})} d\tau\\
&\leq CT^\alpha h^\kappa.
\end{aligned}$$

This also converges to zero, as $h \to 0^+$, uniformly for $t \in [\delta, T-\delta]$. Finally, in order to estimate the summand $II(t,h)$, we note that $w_0 := f(0, u(0)) \in L^p(\mathcal{X})$ (on account of the fact $u_0 = u(0) \in D(A_p) \subset L^\infty(\mathcal{X})$) and that $P_\alpha(t)$ is continuously differentiable for all $t \geq \delta$, which then yields

$$\frac{1}{h}\left\|\int_t^{t+h} P_\alpha(\tau) w_0 ds - P_\alpha(t) w_0\right\|_{L^p(\mathcal{X})} d\tau \to 0, \text{ as } h \to 0^+, \qquad (3.3.18)$$

uniformly for all $t \in [\delta, T - \delta]$. Let $\varepsilon > 0$ be given. We can pick a sufficiently small $h_0 \leq \delta$, such that for $0 < h \leq h_0$, it holds

$$\begin{aligned}
&\|II(t,h)\|_{L^p(\mathcal{X})} \\
&\leq \frac{1}{h}\left\| \int_t^{t+h} P_\alpha(\tau) w_0 - P_\alpha(t) w_0 d\tau \right\|_{L^p(\mathcal{X})} \\
&+ \frac{1}{h}\left\| \int_t^{t+h} P_\alpha(\tau)\left(f(t+h-\tau, u(t+h-\tau)) - f(0, u(0))\right) d\tau \right\|_{L^p(\mathcal{X})} \\
&\leq \varepsilon + C_p h^{-1} \int_t^{t+h} \tau^{\alpha-1}\left[(t+h-\tau) + \|u(t+h-\tau) - u(0)\|_{L^\infty(\mathcal{X})}\right] d\tau \\
&\leq \varepsilon + C_{p,T} h^{-1} \int_t^{t+h} \tau^{\alpha-1}(t+h-\tau)^\kappa \left[1 + (t+h-\tau)^{1-\kappa}\right] d\tau \\
&\leq \varepsilon + C_{p,T,\delta}\left(1 + T^{1-\kappa}\right) h^{-1} \int_t^{t+h} (t+h-\tau)^\kappa \, d\tau \\
&\leq \varepsilon + \frac{C_{p,T,\delta}\left(T^{1-\kappa}+1\right)}{\kappa+1} h^\kappa \\
&\leq 2\varepsilon,
\end{aligned}$$

owing once again to the second of (3.3.1). Here the constant $C_{p,T,\delta} \sim \delta^{\alpha-1}$ occurs as the maximum of the function $\tau \mapsto \tau^{\alpha-1}$ over $[\delta, T]$. Collecting all these previous estimates, we have completed the proof of (i).

We next give a proof of (ii). Recall the definition of $Q(\tau, h)$ from (3.3.11). We begin with the basic estimate for $0 < \tau < t$,

$$\begin{aligned}
\|v(\tau)\|_{L^p(\mathcal{X})} &\leq \left\| S_\alpha'(\tau) u_0 \right\|_{L^p(\mathcal{X})} + \|P_\alpha(\tau) f(0, u(0))\|_{L^p(\mathcal{X})} \\
&+ \int_0^\tau \|P_\alpha(s-\xi)\|_{p,p} \|\partial_t f(\xi, u(\xi))\|_{L^p(\mathcal{X})} d\xi \\
&\leq C_p \tau^{\alpha-1}\left(\left\|A_p u_0\right\|_{L^p(\mathcal{X})} + \|w_0\|_{L^p(\mathcal{X})}\right) + K \cdot C_p \tau^\alpha \\
&\leq C_{p,T,K}(u_0) \tau^{a-1},
\end{aligned} \tag{3.3.19}$$

for some constant $C_{p,T,K} > 0$ independent of h, τ and t. Clearly, whenever $0 < h \leq h_0$ and $0 < \tau < t$,

$$\begin{aligned}
h^{-1}\|(S_\alpha(\tau+h) - S_\alpha(\tau)) u_0\|_{L^p(\mathcal{X})} &\leq 1 + \left\| S_\alpha'(\tau) u_0 \right\|_{L^p(\mathcal{X})} \\
&\leq 1 + C_p \left\| A_p u_0 \right\|_{L^p(\mathcal{X})} \tau^{\alpha-1}.
\end{aligned} \tag{3.3.20}$$

Moreover, it follows that

$$h^{-1}\int_{\tau}^{\tau+h}\|P_{\alpha}(\xi)f(s+h-\xi,u(\tau+h-\xi))\|_{L^p(\mathcal{X})}d\xi \le h^{-1}C_{p,K}\int_{\tau}^{\tau+h}\xi^{\alpha-1}d\xi,$$

with a function on the right hand side that equals $C((\tau+h)^{\alpha}-\tau^{\alpha})h^{-1}$, for some constant $C > 0$ depending only on $C_{p,K}$ and α. We have $(\tau+h)^{\alpha}-\tau^{\alpha} \le h\tau^{\alpha-1}$ for all $h, \tau > 0$ (indeed set $r := h/\tau$ and notice that $(1+r)^{\alpha} \le 1+r$, for any $r > 0$ and $\alpha \in (0,1)$). Then for all $0 < \tau < t$,

$$h^{-1}\left\|\int_{\tau}^{\tau+h}P_{\alpha}(\xi)f(\tau+h-\xi,u(\tau+h-\xi))d\xi\right\|_{L^p(\mathcal{X})} \le C\tau^{\alpha-1}. \tag{3.3.21}$$

Next, it is also obvious that

$$\left\|\int_{0}^{\tau}P_{\alpha}(\xi)\partial_t f(\tau+\theta(h)-\xi,u(\tau-\xi))\right\|_{L^p(\mathcal{X})}d\xi \le C_p\cdot K\tau^{\alpha-1}. \tag{3.3.22}$$

Combining all the estimates from (3.3.19)–(3.3.22), we can write a bound in the form

$$\|R(\tau,h)-v(\tau)\|_{L^p(X)} \le K_0+K_1\tau^{\alpha-1},\ \text{for } 0 < \tau < t. \tag{3.3.23}$$

Finally, for the summand $IV(\tau,h)$ in (3.3.15), whenever $0 < \tau < t < T$, we have

$$\|IV(\tau,h)\|_{L^p(\mathcal{X})} \le 2K\cdot C_p\int_{0}^{\tau}(\tau-\xi)^{\alpha-1}d\xi \le C\tau^{\alpha}. \tag{3.3.24}$$

Collecting the foregoing inequalities (3.3.23)–(3.3.24), we have thus concluded the estimate for $Q(\tau,h)$ in (ii).

We now finalize the proof of the theorem. Set $\zeta := cT^{\alpha}E_{\alpha,\alpha+1}(cT^{\alpha})$ for the same constant $c > 0$ from (3.3.10) and observe also that by (A.0.19), we have that

$$\int_{0}^{T}\tau^{\alpha-1}E_{\alpha,\alpha}\left(c\tau^{\alpha}\right)d\tau = \zeta.$$

Let $K_2 > 0$ be a bound for the function $t \mapsto Ct^{\alpha-1}E_{\alpha,\alpha}(ct^{\alpha})$ over $\delta \le t \le T-\delta$. Given $\varepsilon > 0$, pick $\eta \in (0,\delta]$ sufficiently small so that

$$\int_{0}^{\eta}K_2\left(K_0+K_1\tau^{\alpha-1}\right)d\tau < \varepsilon(2+\zeta)^{-1}.$$

Now pick h_0 so small such that whenever $0 < h \le h_0$, $\|Q(t,h)\|_{L^p(\mathcal{X})} \le \varepsilon(2+\zeta)^{-1}$, uniformly in the range $\delta \le t \le T-\delta$. Then for all $h \in (0,h_0]$

and $t \in [\delta, T-\delta]$ one has from (3.3.10) that

$$\begin{aligned} &\|Z(t,h)-V(t)\|_{L^p(\mathcal{X})} \\ \leq &\|Q(t,h)\|_{L^p(\mathcal{X})} \\ &+C\int_0^\eta (t-\tau)^{\alpha-1} E_{\alpha,\alpha}\left(c(t-\tau)^\alpha\right)\|Q(\tau,h)\|_{L^p(\mathcal{X})}\,d\tau \\ &+C\int_\eta^t (t-\tau)^{\alpha-1} E_{\alpha,\alpha}\left(c(t-\tau)^\alpha\right)\|Q(\tau,h)\|_{L^p(\mathcal{X})}\,d\tau \\ \leq &\varepsilon(2+\zeta)^{-1}+\int_0^\eta K_2\left(K_0+K_1\tau^{\alpha-1}\right)d\tau \\ &+\varepsilon(2+\zeta)^{-1}\int_\eta^t (t-\tau)^{\alpha-1} E_{\alpha,\alpha}\left(c(t-\tau)^\alpha\right)d\tau \\ \leq &2\varepsilon(2+\zeta)^{-1}+\varepsilon(2+\zeta)^{-1}\int_0^T \tau^{\alpha-1}E_{\alpha,\alpha}\left(c\tau^\alpha\right)d\tau \\ \leq &\frac{2\varepsilon}{2+\zeta}+\varepsilon(2+\zeta)^{-1}\zeta=\varepsilon. \end{aligned} \tag{3.3.25}$$

Since $\varepsilon > 0$ is arbitrary, (3.3.25) shows that $Z(t,h) \to V(t)$ as $h \to 0^+$ uniformly in $\delta \leq t \leq T-\delta$. But $\delta > 0$ is also arbitrary so that this argument also gives that $V(t)$ is the continuous right derivative of $u(t)$ on the interval $t \in (0,T)$. Since the convergence $Z(t,h) \to V(t)$ is also uniform on any interval of the form $[\delta, T-\delta]$, the set $\{Z(\cdot,h) : 0 < h < \delta\}$ of $L^p(\mathcal{X})$-valued functions is equicontinuous. Therefore,

$$\lim_{h\to 0^+} Z(t,h) = \lim_{h\to 0^+} Z(t-h,h) = V(t) \tag{3.3.26}$$

uniformly on any interval $t \in [\delta, T-\delta]$ (the second limit in (3.3.26) shows that $V(t)$ is also the continuous left derivative of $u(t)$ on $[\delta, T-\delta]$). But since $\delta > 0$ was arbitrary, it follows that $u'(t) = V(t)$ for all $t \in (0,T)$. Finally, arguing in a similar fashion as above, we also deduce that $V(T)$ is the left-derivative of $u(t)$ at $t = T$, and therefore, the conclusion of the theorem follows. □

Remark 3.3.3 More involved arguments should also give that u is $C^k((0,T]; L^p(\mathcal{X}))$ for any integer $k \geq 1$, provided that the nonlinear function f is sufficiently smooth (see Chap. 5, Problem 4).

We conclude the section by formulating the following fundamental question.

Under what conditions can we say that the notions of generalized Caputo derivative (see Definition 2.1.1) and the classical Caputo derivative $g_{1-\alpha} * \partial_t f$ *are equivalent for a solution of problem (3.1.1)?*

A classical result states this equivalency, for every $t \in [0, T]$, for functions $u : [0, T] \to Y$ that are differentiable everywhere on $[0, T]$ (see (2.1.3)). However, any strong solution $u : [0, T] \to Y = L^p(\mathcal{X})$, of Eq. (3.1.1), even in the homogeneous linear case of (2.1.6), is *not* differentiable at $t = 0$, no matter how smooth the initial condition is.[2] It turns out that a more general result that does *not* require (classical) differentiability at $t = 0$ can hold based on [8, Theorem 3.1], namely, $\partial_t^\alpha u(t) = (g_{1-\alpha} * \partial_t u)(t)$, for *almost all* $t \in (0, T]$, for as long as the function $u \in AC([0, T]; Y)$ and Y is Gelfand (i.e., Y has the Radon-Nikodym property with respect to the Lebesgue measure on the Borel sets of $[0, T]$). Here $AC([0, T]; Y)$ denotes the space of all absolutely continuous functions on $[0, T]$ with values in the Banach space Y, namely, for each $\varepsilon > 0$ there exists a $\delta > 0$ such that if (a_n, b_n) is a sequence of disjoint subintervals of $[0, T]$, with $\sum_n (b_n - a_n) < \delta$, then

$$\sum_n \|u(b_n) - u(a_n)\|_Y < \varepsilon.$$

The space $Y = L^p(\mathcal{X})$, $1 < p < \infty$, is an example of such a Gelfand-space (see [7, Chapter IV, Section 3 and Chapter V]). In this case, it also follows directly from the Radon-Nikodym theorem (see again [7]) that

$$AC([0, T]; L^p(\mathcal{X})) \cong W^{1,1}([0, T]; L^p(\mathcal{X})),$$

provided that $\mathcal{X}$ satisfies all the assumptions of Chap. 2.

The next result provided us with the first concrete evidence of the validity of the equivalency problem for any strong solution of problem (3.1.1).

Theorem 3.3.4 *Let the assumptions of Theorem 3.3.2 be satisfied. Then every strong solution on* $[0, T]$ *(with* $T < T_{\max}$*), of problem (3.1.1), also satisfies the initial value problem*

$$g_{1-\alpha} * \partial_t u = Au + f(x, t, u), \text{ for almost all } \mathcal{X} \times (0, T], \ u(\cdot, 0) = u_0 \text{ in } \mathcal{X}. \tag{3.3.27}$$

Proof The proof is an immediate consequence of the proof of Theorem 3.3.2; it is an easy exercise since

$$u \in C^1((0, T]; L^p(\mathcal{X})) \cap C^{0,\kappa}([0, T]; L^\infty(\mathcal{X})), \ 0 < \kappa < 1.$$

Indeed, it follows that $\|\partial_t u(t)\|_{L^p(\mathcal{X})} \leq C_{p,T} t^{\alpha-1}$, for all $t \in (0, T]$, and therefore, it holds

$$u \in W^{1,q}([0, T]; L^p(\mathcal{X})) \subseteq W^{1,1}([0, T]; L^p(\mathcal{X})),$$

for any $1 \leq q < 1/(1-\alpha)$. Based on the prior statements above, the proof then follows. □

---

[2] This is in contrast to the case $\alpha = 1$.

3.4 Global A Priori Estimates

Our goal in this subsection is to derive an ***explicit*** uniform L^{∞}-estimate from some given L^{r}-estimate of the mild (or quasi-mild) solution. Indeed, according to previous results, the L^{∞}-bound is crucial for the global regularity problem as well as it is essential in the investigation of the long-time behavior as time goes to infinity (see, for instance, [11–13]). In what follows we shall implicitly make use of the fact that every mild solution constructed in this section is in fact a strong solution on some maximal interval of existence. In the case $u_0 \in \mathcal{L}^{\infty}(\mathcal{X})$, this statement is already a consequence of Theorem 3.2.6 and the local boundedness of the mild solution (see Theorem 3.1.9). In the case when $u_0 \in L^{p_0}(\mathcal{X})$, $p_0 \in [1, \infty)$ the arguments below can still be made rigorous by employing a regularization procedure in which $u_{0j} \in D(A_p) \subset L^{\infty}(\mathcal{X})$, for $p \geq p_0$ with $p \in (\beta_A, \infty) \cap (1, \infty)$, such that $u_{0j} \to u_0$ in $L^p(\mathcal{X})$ (since $D(A_p)$ is dense in $L^p(\mathcal{X})$). This is no serious drawback since the corresponding mild solutions associated with the initial datum u_{0j} are indeed strong solutions and every mild solution associated with the initial datum u_0 is locally bounded (see again Theorem 3.1.9).

We shall present two methods which are of different nature. First, we appeal to a method exploited in [22] for classical systems of parabolic equations ($\alpha = 1$) with standard diffusion Δ, and which is based on **"feedback"** and some **bootstrap** arguments. The advantage of the "feedback" argument is that it uses only elementary inequalities. The next theorem generalizes the "feedback" argument to nonlocal problems with a fractional-in-time Caputo derivative and a larger class of "diffusion" operators A, possibly also nonlocal (see Sect. 2.3). Besides, it also has direct application in the theory dealing with nonlinear systems of fractional kinetic equations. We refer the reader to Chap. 4 below and the corresponding sections.

As before, we recall that $q_1 \in (\beta_A, \infty] \cap [1, \infty]$ and $q_2 \in (1/\alpha, \infty]$, and

$$\mathfrak{n} := \beta_A \alpha > 0,\ 0 \leq \delta := \frac{\mathfrak{n}}{p_0} - \frac{\mathfrak{n}}{p} < \alpha$$

where $\beta_A > 0$, $0 < \alpha \leq 1$, for all $p \in [p_0, \infty]$. Let

$$\Upsilon(t) := \begin{cases} t^{\varepsilon}, & \text{if } t \in (0, 1] \\ t, & \text{if } t > 1. \end{cases}$$

We note that if $p_0 \geq \max(\beta_A, 1)$ we automatically have $\delta \in [0, \alpha)$, for arbitrary $p \in [p_0, \infty]$ while for $p_0 < \beta_A$ this holds provided that $p_0 \leq p < p_0/(1 - p_0/\beta_A)$. Having the restriction $\delta \in [0, \alpha)$ is required only in the case $\alpha \in (0, 1)$ due to (2.2.12). When $\alpha = 1$, such a restriction can be eliminated and it suffices to ask instead that $\delta \geq 0$. Further conditions on $p \in [p_0, \infty]$ will be sought in the proofs of the supporting lemmatas below.

Theorem 3.4.1 (Global A Priori Estimate) *Let $r_1, r_2 \in [1, \infty]$, q_1, q_2 and $\gamma \in [1, \infty)$ satisfy*

$$\frac{n}{q_1} + \frac{1}{q_2} + (\gamma - 1)\left(\frac{n}{r_1} + \frac{1}{r_2}\right) < \alpha \text{ and } \frac{1}{q_1} + \frac{\gamma - 1}{r_1} < 1, \text{ if } r_1 < \infty. \tag{3.4.1}$$

Let $p_0 \in [1, \infty]$ be arbitrary and assume f obeys the conditions **(F1)–(F2)** *for some $\gamma \in [1, \infty)$ and q_1, q_2 satisfying (3.4.1). Let now $u_0 \in \mathcal{L}^{p_0}(\mathcal{X}) \subseteq L^{p_0}(\mathcal{X})$ ($p_0 \in [1, \infty]$) for which the corresponding mild solution satisfies*

$$\|u\|_{r_1, r_2, T} \le L(\|u_0\|_{L^{p_0}(\mathcal{X})}) \tag{3.4.2}$$

on any time interval $[0, T]$, for some positive increasing function L (independent of u, u_0) but which depends on the $L^{p_0}(\mathcal{X})$-norm of u_0. Then Problem (3.1.1) has a unique global mild (or quasi-mild, if $p_0 = \infty$) solution on $[0, \infty)$ in the sense of Definition 3.1.1 (and Definition 3.1.3, respectively). In particular, there exist numbers $\rho > 0$ and $\varepsilon > 0$ such that the mild solution u satisfies the estimates:

$$\sup_{t \in (0,\infty)} (t \wedge 1)^{\delta} \|u(\cdot, t)\|_{L^p(\mathcal{X})} < \infty, \text{ for all } p \in [p_0, \infty] \tag{3.4.3}$$

and

$$\|u(\cdot, t)\|_{L^\infty(\mathcal{X})} \le C (t \wedge 1)^{-\frac{n}{p_0}} \left[\|u_0\|_{L^{p_0}(\mathcal{X})} + \Upsilon(t)\left(\Phi + \Phi^{\rho}\right)\right], \tag{3.4.4}$$

where $\Phi = \Phi\left(\|u_0\|_{L^{p_0}(\mathcal{X})}\right) := \left(1 + L\left(\|u_0\|_{L^{p_0}(\mathcal{X})}\right)\right)(1 + \|c\|_{q_1,q_2})$. Estimate (3.4.4) holds with $\rho = \gamma$ if one assumes that

$$\frac{n}{q_1} + \frac{1}{q_2} + \gamma\left(\frac{n}{r_1} + \frac{1}{r_2}\right) < \alpha \text{ and } \frac{1}{q_1} + \frac{\gamma}{r_1} \le 1.$$

The proof of this theorem is based on some subsequent lemmas. Note that the next lemma does not use the Lipschitz condition (**F2**).

Lemma 3.4.2 *Let $p_0 \in [1, \infty]$, $r_1, r_2 \in [1, \infty]$, $\gamma \in [1, \infty)$, $b \in [0, 1]$, $\varepsilon \in (0, \alpha)$ and $p \in [p_0, \infty]$ such that*

$$\begin{cases} \dfrac{n}{q_1} + \dfrac{1}{q_2} + \gamma(1 - b)\left(\dfrac{n}{r_1} + \dfrac{1}{r_2}\right) < \alpha + (1 - \gamma b)\dfrac{n}{p} - \varepsilon, \\ \dfrac{1}{q_1} + \gamma\dfrac{(1 - b)}{r_1} + \gamma\dfrac{b}{p} \le 1, \\ \dfrac{1}{q_2} + \gamma\dfrac{(1 - b)}{r_2} + \gamma b\delta < \alpha - \varepsilon, \\ \gamma b < 1, \text{ with } \delta = \dfrac{n}{p_0} - \dfrac{n}{p} \in [0, \alpha). \end{cases} \tag{3.4.5}$$

Let f obey the condition **(F1)** *for some $\gamma \in [1, \infty)$ satisfying (3.4.5). Let the mild solution u for Problem (3.1.1) with an initial datum $u_0 \in \mathcal{L}^{p_0}(\mathcal{X}) \subset L^{p_0}(\mathcal{X})$ satisfy the a priori estimate $\|u\|_{r_1,r_2,T} < \infty$, for any $T > 0$. Furthermore assume $|||u|||_{p,\delta,T} < \infty$ and for $b > 0$ define*

$$U := \|1 + |u|\|_{r_1,r_2,T}^{\gamma(1-b)} \|c\|_{q_1,q_2} .$$

Then there exists a constant $C_* > 0$ independent of u_0, u, U, t and T such that

$$|||u|||_{p,\delta,T} \leq C_* \left[\|u_0\|_{L^{p_0}(\mathcal{X})} + \Upsilon(T) \left(U + U^{1/(1-\gamma b)}\right) \right]. \tag{3.4.6}$$

Proof We shall exploit again the integral formulation for the mild solution (see Definition 3.1.1). By (3.4.5), there exist $s_1, s_2 \in [1, \infty]$ such that

$$\frac{n}{s_1} + \frac{1}{s_2} \leq \alpha + \frac{n}{p} - \varepsilon, \tag{3.4.7}$$

$$\frac{1}{s_2} + \gamma b \delta < \alpha - \varepsilon \leq 1 - \varepsilon, \tag{3.4.8}$$

$$\frac{1}{q_1} + \frac{\gamma(1-b)}{r_1} + \frac{\gamma b}{p} \leq \frac{1}{s_1}, \tag{3.4.9}$$

$$\frac{1}{q_2} + \frac{\gamma(1-b)}{r_2} \leq \frac{1}{s_2}. \tag{3.4.10}$$

We have

$$\|u(\cdot,t)\|_{L^p(\mathcal{X})} \leq \|S_\alpha(t)\|_{p,p_0} \|u_0\|_{L^{p_0}(\mathcal{X})} + \int_0^t \|P_\alpha(t-\tau)\|_{p,s_1} \|f(\tau, u(\cdot,\tau))\|_{L^{s_1}(\mathcal{X})} d\tau.$$

We use (**F1**), to split the nonlinear term into several terms. First, by (3.4.9) and the Hölder inequality we get for all $t > 0$,

$$\begin{aligned} &(t \wedge 1)^\delta \|u(\cdot,t)\|_{L^p(\mathcal{X})} \\ &\leq (t \wedge 1)^\delta \|S_\alpha(t)\|_{p,p_0} \|u_0\|_{L^{p_0}(\mathcal{X})} \\ &\quad + (t \wedge 1)^\delta \int_0^t \|P_\alpha(t-\tau)\|_{p,s_1} \|c(\cdot,\tau)\|_{L^{q_1}(\mathcal{X})} \|1 + |u(\tau)|\|_{L^{r_1}(\mathcal{X})}^{\gamma(1-b)} (\tau \wedge 1)^{-\gamma b \delta} d\tau \\ &\quad \times \left(|||1 + |u|||_{p,\delta,T}^{\gamma b} \right). \end{aligned} \tag{3.4.11}$$

The first summand on the right-hand side of (3.4.11) can be estimated using the ultracontractivity property (2.2.12) for $S_\alpha(t)$ as a bounded operator from $L^{p_0}(\mathcal{X})$ into $L^p(\mathcal{X})$. For the second summand we apply Lemma A.0.1 with the choice

$p, s_1, s_2, \delta, \varepsilon$ as above, $\theta := \gamma b\delta$ and $r(\tau) := \|c(\cdot,\tau)\|_{L^{q_1}(\mathcal{X})} \|1 + |u(\cdot,\tau)|\|_{L^{r_1}(\mathcal{X})}^{\gamma(1-b)}$. Hence from (3.4.11), we deduce

$$|||u|||_{p,\delta,T} \le C \|u_0\|_{L^{p_0}(\mathcal{X})} + C\Upsilon(T) \, p_{s_2}(r) \, |||1 + |u|||\,|_{p,\delta,T}^{\gamma b}. \tag{3.4.12}$$

The function $p_{s_2}(r)$ can be estimated from the same Lemma A.0.1 using the Hölder inequality on account of (3.4.10). It follows that $p_{s_2}(r) \le \|c\|_{q_1,q_2} \|1 + |u|\|_{r_1,r_2,T}^{\gamma(1-b)} = U$. Therefore, (3.4.12) implies that

$$|||u|||_{p,\delta,T} \le C \|u_0\|_{L^{p_0}(\mathcal{X})} + C\Upsilon(T) \, U \, |||1 + |u|||\,|_{p,\delta,T}^{\gamma b}. \tag{3.4.13}$$

Observe now that (3.4.13) is already the assertion (3.4.6) when $b = 0$. In order to show the estimate in the case when $b > 0$, we apply a "feedback" argument to (3.4.13) by employing the "feedback" inequality of Lemma A.0.6 with the following choices

$$y := |||u|||_{p,\delta,T}, \quad z_0 := C\left(\|u_0\|_{L^{p_0}(\mathcal{X})} + \Upsilon(T)\, U\right), \quad z_1 := C\Upsilon(T)\, U$$

with $\sigma := \gamma b < 1$. Indeed, (3.4.13) yields that $y \le z_0 + z_1 y^{\sigma}$ and therefore, we obtain

$$y \le \frac{z_0}{1-\sigma} + z_1^{\frac{1}{1-\sigma}}.$$

The foregoing inequality yields (3.4.6) with constant $C_* = C/(1-\gamma b) + C^{1/(1-\gamma b)}$.

Next, we can also check in what sense the initial datum is satisfied. By the integral formula and the bound (**F1**) we have

$$\|u(\cdot,t) - S_\alpha(t) u_0\|_{L^{p_0}(\mathcal{X})}$$

$$\le \int_0^t \|P_\alpha(t-\tau)\|_{p_0,s_1} \|c(\cdot,\tau)\|_{L^{q_1}(\mathcal{X})} \|1 + |u(\cdot,\tau)|\|_{L^{r_1}(\mathcal{X})}^{\gamma(1-b)} \|1 + |u(\cdot,\tau)|\|_{L^{r_1}(\mathcal{X})}^{\gamma b} \, d\tau$$

on which we can once again apply Lemma A.0.1 with the same s_1, s_2 and choice of function $r(s)$ as above, and $\delta := 0$, $p := p_0$, $\theta := \gamma b\delta$ and $\varepsilon := \varepsilon$. By (3.4.7) and (3.4.8), we can easily verify that the assumptions of Lemma A.0.1 are indeed verified. We get

$$\begin{aligned} \|u(\cdot,t) - S_\alpha(t) u_0\|_{L^{p_0}(\mathcal{X})} &\le C\Upsilon(t) \, p_{s_2}(r) \, |||1 + |u|||\,|_{p,\delta,T}^{\gamma b} \\ &\le C\Upsilon(t) \, U \, |||1 + |u|||\,|_{p,\delta,T}^{\gamma b}, \end{aligned} \tag{3.4.14}$$

for all $t \in (0, T]$. Notice that $\Upsilon(t) \to 0$ as $t \to 0^+$. Finally, it is also easy to check that $\|f(s, u(s))\|_{1,1,T} < \infty$, for any $T > 0$ for which u satisfies (3.4.6). Indeed, we get

$$\|f(s, u(s))\|_{1,1,T} \le C(T + \Upsilon(T))\, U\, |||1 + |u|||_{p,\delta,T}^{\gamma b} \overset{(3.4.6)}{<} \infty .$$

The proof is finished. □

Note also that the next supporting result does not use the Lipschitz condition **(F2)**.

Lemma 3.4.3 *Let q_1, q_2, and $r_1, r_2 \in [1, \infty]$, $\gamma \in [1, \infty)$, $b \in [0, 1]$, $\varepsilon \in (0, \alpha)$ satisfy*

$$\frac{\mathfrak{n}}{q_1} + \frac{1}{q_2} + \gamma(1 - b)\left(\frac{\mathfrak{n}}{r_1} + \frac{1}{r_2}\right) < \alpha - \varepsilon, \tag{3.4.15}$$

$$\frac{1}{q_1} + \frac{\gamma(1 - b)}{r_1} \le 1, \tag{3.4.16}$$

$$\gamma b < 1. \tag{3.4.17}$$

Let $p_0 \in [1, \infty]$ be arbitrary and let f satisfy condition **(F1)** *for some $\gamma \in [1, \infty)$ that obeys* (3.4.15)–(3.4.17)*. Let the mild solution u for problem* (3.1.1) *with an initial datum $u_0 \in \mathcal{L}^{p_0}(\mathcal{X})$ satisfy $U < \infty$ for any $T > 0$, where U is defined in the statement of Lemma 3.4.2. Furthermore for $b > 0$ assume that $\sup_{t \in (0,T]} \|u(\cdot, t)\|_{L^\infty(\mathcal{X})} < \infty$. Then there exists a constant $C_* > 0$ independent of u_0, u, U, t and T such that*

$$\|u(\cdot, t)\|_{L^\infty(\mathcal{X})} \le C_*(t \wedge 1)^{-\frac{\mathfrak{n}}{p_0}} \left[\|u_0\|_{L^{p_0}(\mathcal{X})} + \Upsilon(t)\left(U + U^{1/(1-\gamma b)}\right)\right], \tag{3.4.18}$$

for all $t \in (0, T]$.

Remark 3.4.4 In the case $b > 0$ the a priori information $\sup_{t \in (0,T]} \|u(\cdot, t)\|_{L^\infty(\mathcal{X})} < \infty$ is essential to deduce the explicit estimate (3.4.18) with a constant independent of time and of any $T > 0$. Otherwise, no conclusion can be drawn from the "feedback" argument. On the other hand since every mild solution of Theorem 3.1.4 is locally bounded by Theorem 3.1.9 on $(0, T_{\max})$, we can infer from (3.4.18) that $T_{\max} = \infty$ for as long as U is finite on any time interval $[0, T]$.

Proof (Proof of Lemma 3.4.3) First, we observe that when $p_0 = \infty$ or $b = 0$, the assumptions (3.4.5) of Lemma 3.4.2 are satisfied with $p := \infty$, $\delta := 0$ and r_1, r_2, γ as above in (3.4.15)–(3.4.17). In this case, the assertion (3.4.18) is equivalent to the estimate (3.4.6) of Lemma 3.4.2. Thus, we may assume that $p_0 \in [1, \infty)$ and $b \in (0, 1]$. We apply an inductive argument with help from Lemmas 3.4.2 and A.0.5

in the appendix. To this end, consider the finite sequences $\{p_i\}$ with $p_0 < p_1 < \ldots < p_k = \infty$, and $\{\delta_i\} \in (0, \alpha)$ for $i = 1, .., k$ as given by Lemma A.0.5. We then apply Lemma 3.4.2 with the choices $p_0 := p_{i-1}$, $p := p_i$, $\delta := \delta_i$, the exponents r_1, r_2, γ, b as above in (3.4.15)–(3.4.17), and initial datum $u_0 := u(t)$ for arbitrary $t \in (0, T]$. It follows that (3.4.6) of Lemma 3.4.2 yields for all $h \in (0, T - t]$, $i = 1, .., k$,

$$|||u(t+h)|||_{p_i,\delta_i,h} \leq C_i \left[\|u(\cdot, t)\|_{L^{p_{i-1}}(\mathcal{X})} + \Upsilon(h)\left(U + U^{1/(1-\gamma b)}\right)\right]. \tag{3.4.19}$$

The choice $t = ih - h$ in (3.4.19) then gives

$$\begin{aligned} &(h \wedge 1)^{\delta_i} \|u(\cdot, ih)\|_{L^{p_i}(\mathcal{X})} \\ &\leq C_i \left[\|u(\cdot, ih - h)\|_{L^{p_{i-1}}(\mathcal{X})} + \Upsilon(h)\left(U + U^{1/(1-\gamma b)}\right)\right], \end{aligned} \tag{3.4.20}$$

for all $i = 1, .., k$ and $h \in (0, T/i]$, for some $C_i < \infty$. An induction argument in (3.4.20) for $i = 1, \ldots, k$ implies

$$(h \wedge 1)^{\delta_1+\ldots+\delta_i} \|u(\cdot, ih)\|_{L^{p_i}(\mathcal{X})} \leq C_i \left[\|u_0\|_{L^{p_0}(\mathcal{X})} + \Upsilon(h)\left(U + U^{1/(1-\gamma b)}\right)\right]. \tag{3.4.21}$$

Since $\delta_i = \mathfrak{n}/p_{i-1} - \mathfrak{n}/p_i$, we readily have $\delta_1 + \ldots + \delta_k = \mathfrak{n}/p_0 - \mathfrak{n}/p_k = \mathfrak{n}/p_0$ and (3.4.21) with $i = k$, gives no other than the required estimate (3.4.18). This completes the proof of the lemma. □

Remark 3.4.5 The statements of Lemmas 3.4.2 and 3.4.3 reduce exactly to those of [22, Lemma 19 and Lemma 20] in the case $\alpha = 1$ and $A = \Delta$ (with $\beta_A = N/2$). Finally, we also observe that when $u_0 \geq 0$ and $u \geq 0$, the assumption **(F1)** in these lemmas can be replaced by the weaker one-sided bound $f(x, t, u) \leq c(x, t)(1 + u^\gamma)$. Indeed, since both families $\{S_\alpha(t)\}$, $\{P_\alpha(t)\}$ are positive by Proposition 2.2.1, the following pointwise estimate holds

$$0 \leq u(t) \leq v(t) := S_\alpha(t) u_0 + \int_0^t P_\alpha(t-\tau)\left(c(\cdot, \tau)\left(1 + u^\gamma(\tau)\right)\right) d\tau,$$

and the whole arguments of Lemmas 3.4.2 and 3.4.3 can be instead applied to the integral family v.

Before we can finish the proof of Theorem 3.4.1 we also need the following continuous dependence estimate. This result uses the Lipschitz condition **(F2)** in a crucial way.

Lemma 3.4.6 *Let $p_0 \in [1,\infty]$, $r_1, r_2 \in [1,\infty]$, $\varepsilon \in (0,\alpha)$, $p \in [p_0,\infty]$ and assume* **(F2)** *for some $\gamma \in [1,\infty)$, q_1, q_2 that satisfy*

$$\frac{\mathfrak{n}}{q_1} + \frac{1}{q_2} + (\gamma - 1)\left(\frac{\mathfrak{n}}{r_1} + \frac{1}{r_2}\right) < \alpha - \varepsilon, \tag{3.4.22}$$

$$\frac{1}{q_1} + \frac{\gamma - 1}{r_1} + \frac{1}{p} \leq 1, \tag{3.4.23}$$

$$\frac{1}{q_2} + \frac{\gamma - 1}{r_2} + \delta < \alpha - \varepsilon, \tag{3.4.24}$$

and the a priori estimate (3.4.2). Let u_i be any two mild solutions in the sense of Definition 3.1.1 for any two initial data $u_{0i} \in \mathcal{L}^{p_0}(\mathcal{X}) \subseteq L^{p_0}(\mathcal{X})$, $i = 1, 2$. Then there exists a constant $C > 0$ independent of u_i, t, T and u_{0i}, such that

$$\begin{aligned} |||u_1 - u_2|||_{p,\delta,T} \leq & C \|u_{01} - u_{02}\|_{L^{p_0}(\mathcal{X})} \\ & + C\Upsilon(T) \|1 + |u_1| + |u_2|\|_{r_1,r_2,T}^{\gamma-1} |||u_1 - u_2|||_{p,\delta,T}, \end{aligned} \tag{3.4.25}$$

for all $t \in (0, T]$.

Proof The argument follows in a similar fashion to the computation (3.4.12)–(3.4.13) using the local Lipschitz condition (**F2**). Choose $s_1, s_2 \in [1,\infty]$ such that

$$\frac{1}{q_1} + \frac{\gamma - 1}{r_1} + \frac{1}{p} \leq \frac{1}{s_1}, \qquad \frac{1}{q_2} + \frac{\gamma - 1}{r_2} \leq \frac{1}{s_2}$$

and

$$\frac{\mathfrak{n}}{s_1} + \frac{1}{s_2} - \frac{\mathfrak{n}}{p} + \varepsilon < \alpha, \qquad \frac{1}{s_2} + \delta \leq \alpha - \varepsilon.$$

By the integral solution representation for each u_i, by the Hölder inequality and (**F2**) we have

$$\begin{aligned} & \|(u_1 - u_2)(\cdot, t)\|_{L^p(\mathcal{X})} \\ & \leq \|S_\alpha(t)\|_{p,p_0} \|u_{01} - u_{02}\|_{L^{p_0}(\mathcal{X})} \\ & \quad + |||u_1 - u_2|||_{p,\delta,T} \\ & \times \int_0^t \|P_\alpha(t-\tau)\|_{p,s_1} \|c(\cdot,\tau)\|_{L^{q_1}(\mathcal{X})} \|1 + |u_1(\cdot,\tau)| + |u_2(\cdot,\tau)|\|_{L^{r_1}(\mathcal{X})}^{\gamma-1} (\tau \wedge 1)^{-\delta} \, d\tau. \end{aligned} \tag{3.4.26}$$

The first term at the right hand side of (3.4.26) can be estimated as before using the ultracontractivity estimate for $S_\alpha(t)$. For the second summand in (3.4.26), we apply

Lemma A.0.1 (whose assumptions are satisfied) with the choices

$$r(\tau) := \|c(\cdot,\tau)\|_{L^{q_1}(\mathcal{X})} \|1 + |u_1(\cdot,\tau)| + |u_2(\cdot,\tau)|\|^{\gamma-1}_{L^{r_1}(\mathcal{X})}$$

and p, s_1, s_2, ε as above, and $\theta := \delta$. The foregoing inequality then yields

$$|||u_1 - u_2|||_{p,\delta,T} \leq C \|u_{01} - u_{02}\|_{L^{p_0}(\mathcal{X})} + C\Upsilon(T)\, p_{s_2}(r)\, |||u_1 - u_2|||_{p,\delta,T} \cdot \tag{3.4.27}$$

The functional $p_{s_2}(r)$ can be estimated exploiting Lemma A.0.1 once more to find

$$p_{s_2}(r) \leq \|c\|_{q_1,q_2} \|1 + |u_1| + |u_2|\|^{\gamma-1}_{r_1,r_2,T} < \infty,$$

which is finite by virtue of the assumption (3.4.2). Thus, (3.4.27) implies the desired assertion (3.4.25) of Lemma 3.4.6. □

Proof (Proof of Theorem 3.4.1) Let $u_0 \in \mathcal{L}^{p_0}(\mathcal{X})$ and consider a sequence $\{u_{0j}\}_{j\in\mathbb{N}} \subset D(A_p) \subset L^\infty(\mathcal{X})$ for $p \geq p_0$, $p \in (\beta_A, \infty)$, such that

$$\lim_{j\to\infty} \|u_{0j} - u_0\|_{L^{p_0}(\mathcal{X})} = 0 \tag{3.4.28}$$

(recall that $D(A_p)$ is dense in $\mathcal{L}^{p_0}(\mathcal{X})$). By Theorem 3.1.4 there exists a unique mild solution u_j for problem (3.1.1), which is also smooth by Theorem 3.2.2, on the time interval $[0, T_j)$, where $T_j > 0$ is the maximal existence time. We can show that $T_j = \infty$, for all $j \in \mathbb{N}$. The assumption (3.4.1) of Theorem 3.4.1 implies that there exist numbers $\varepsilon \in (0,\alpha)$ and $b \in [0, 1/\gamma)$ such that the assumptions (3.4.15)–(3.4.17) of Lemma 3.4.3 are satisfied. Then we can infer from the estimate (3.4.18) that

$$\|u_j(\cdot,t)\|_{L^\infty(\mathcal{X})} \leq C_*(t \wedge 1)^{-\frac{n}{p_0}} \left[\|u_0\|_{L^{p_0}(\mathcal{X})} + \Upsilon(t)\left(U + U^{1/(1-\gamma b)}\right)\right], \tag{3.4.29}$$

for all $j \in \mathbb{N}$, and $t \in (0, T_j)$. The constant $C_* > 0$ is clearly independent of j. The assertion (3.1.31) together with (3.4.29) and the fact that

$$U = \sup_{j\in\mathbb{N}} \left\{\|c\|_{q_1,q_2}\left(\left\|1 + |u_j|\right\|^{\gamma(1-b)}_{r_1,r_2,T}\right) : T \in (0,\infty)\right\} < \infty$$

uniformly in j, owing to condition (3.4.2), shows that $T_j = \infty$ for all $j \in \mathbb{N}$.

The final goal of the proof is to show, along a proper subsequence (still denoted by) $\{u_j\}$, that u_j converges to a function u on any interval $(0, T] \subset (0,\infty)$. To this end, we also observe that due to the uniform estimate (3.4.29) and the assumption (3.4.2), we have

$$V := \sup_{j,m\in\mathbb{N}} \left\{ \|c\|_{q_1,q_2} \left(\left\| 1 + |u_j| + |u_m| \right\|_{r_1,r_2,T}^{\gamma-1} \right) : T \in (0,\infty) \right\} \tag{3.4.30}$$
$$\leq \|c\|_{q_1,q_2} \left(1 + 2L \left(\|u_0\|_{L^{p_0}(\mathcal{X})} \right) \right)^{\gamma(1-b)}.$$

We choose the initial time ih for an arbitrary $h \in (0,\infty)$ and $i \in \mathbb{N}_0 := \mathbb{N} \cup \{0\}$. The continuous dependence estimate (3.4.25) yields in light of the uniform bound (3.4.30) that

$$\left|\left|\left| \left(u_j - u_l\right)(ih+\cdot) \right|\right|\right|_{p,\delta,h} \tag{3.4.31}$$
$$\leq C \left|\left| \left(u_j - u_l\right)(\cdot, ih) \right|\right|_{L^{p_0}(\mathcal{X})} + C\Upsilon(h)\, V \left|\left|\left| \left(u_j - u_l\right)(ih+\cdot) \right|\right|\right|_{p,\delta,h}^{\gamma b},$$

for all $j,l \in \mathbb{N}$, and $i \in \mathbb{N}_0$ and $h > 0$. Choosing $h \ll 1$ small enough such that $C\Upsilon(h)\, V = Ch^{\varepsilon} V \leq 1/2$, from (3.4.31) we get for all $i \in \mathbb{N}_0$ that

$$\left|\left|\left| \left(u_j - u_l\right)(ih+\cdot) \right|\right|\right|_{p,\delta,h} \leq C \left|\left| \left(u_j - u_l\right)(\cdot, ih) \right|\right|_{L^{p_0}(\mathcal{X})}.$$

In particular, owing to (3.4.28) and a continuation argument, we obtain that $\{u_j\}_{j\in\mathbb{N}}$ is a Cauchy sequence in the Banach space $E_{p,\delta,T}$, for all $T \in (0,\infty)$. Therefore there exists a function $u \in E_{p,\delta,T}$, for any $T \in (0,\infty)$, such that

$$\lim_{j\to\infty} \left|\left|\left| u_j - u \right|\right|\right|_{p,\delta,T} = 0, \text{ for all } T \in (0,\infty). \tag{3.4.32}$$

Fixing now a time $t \in (0,T] \subset (0,\infty)$ and recalling that $\mathcal{X}$ is relatively compact and Hausdorff, (3.4.32) also yields that $u_j(x,t) \to u(x,t)$ (at least for a subsequence) for almost all $x \in \mathcal{X}$; we conclude that (3.4.29) also holds for $u(\cdot,t)$ (as well as the estimate (3.4.3) is verified). Thus, u is well-defined globally on $(0,\infty)$. In order to show that the limit solution u is also a mild solution in the sense of Definition 3.1.1, we argue exactly as in the proof of Theorem 3.1.4, by taking advantage of the strong convergence (3.4.32) to pass to the limit in the integral solution representation for u_j. We leave the simple details to the interested reader.

□

As a consequence of Theorem 3.4.1 we have the following.

Corollary 3.4.7 *Under the assumptions of Theorems 3.1.4 and 3.4.1, every mild* (globally-defined) *solution of Problem* (3.1.1) *with initial datum* $u_0 \in \mathcal{L}^{p_0}(\mathcal{X})$, $p_0 \in [1,\infty]$, *is also a strong solution on* $[T_0,\infty)$, *for any* $T_0 > 0$ *in the sense of Definition 3.2.1.*

A second method to derive global a priori L^{∞}-estimate is based on an iterative Moser procedure used by Alikakos to treat semilinear parabolic problems with diffusion operators that are uniformly elliptic and second-order (see [1]). In this method, the diffusion equation (3.1.1) for $\alpha = 1$ is tested by powers of the solution. Then, estimates of the norms $\|u\|_{L^{2k}}$, are given for $k \in \mathbb{N}$, successively at each step. In [1], each step uses the Gagliardo-Nirenberg inequality but this is not always required (cf. [11–13]). One can cut down on the use of interpolation inequalities by employing instead some Poincaré-Young inequality associated with the corresponding diffusion operator. The latter turns out to be of advantage in the treatment of reaction–diffusion equations ($\alpha = 1$) with "rough" data [12], containing possibly some fractional kinetics [11, 13] (cf. also [10]), when no such interpolation inequalities are in fact available.

Our final goal in this section is to generalize the Moser-Alikakos procedure for the reaction–diffusion equation (3.1.1) in the case $\alpha \in (0, 1)$, that contain all the above studied cases and much more. It is worth mentioning that a priori L^∞-bounds were also recently obtained in [24] for a special case of (3.1.1) with $\alpha \in (0, 1)$, that include quasilinear equations of second order, by exploiting the De Giorgi's iteration technique and suitable truncated energy estimates for weak solutions. Although more general kernels were used, we note that the a priori $L^\infty((0, T) \times \mathcal{X})$-estimates (as obtained by [24]) are weaker than the ones contained here, due to the nature of assumptions contained therein, as well as the following basic observation.

Remark 3.4.8 We have $L^\infty((0, T); L^\infty(\mathcal{X})) \subsetneqq L^\infty((0, T) \times \mathcal{X})$ with strict inclusion (see [9, Chapter 12, Section 2]); the latter space is equipped with norm $\sup_{(t,x)\in(0,T)\times\mathcal{X}} |u(x, t)|$. Note that the Moser procedure always produces a bound in $L^\infty((0, T); L^\infty(\mathcal{X}))$, which is equipped with norm $\sup_{t\in(0,T)} \|u(t)\|_{L^\infty(\mathcal{X})}$.

We first recall an important L^p-norm inequality for the fractional-order derivative ∂_t^α, which has been established recently in [18] (cf. also [25]).

Proposition 3.4.9 *Let the assumptions of Theorem 3.2.6 be satisfied and let $H \in C^1(\mathbb{R})$ be a convex function. Then, for any bounded mild (or quasi-mild) solution of problem* (3.1.1)*, the following inequality holds:*

$$H'(u(t))\, \partial_t^\alpha u(t) \geq \partial_t^\alpha (H(u(t))) = \frac{d}{dt}\Big(g_{1-\alpha} * [H(u(t)) - H(u_0)]\Big)(t),$$

for all $t \in (0, T_{\max})$ and almost all $x \in \mathcal{X}$. In particular, for $H(y) = (1/p)|y|^p$, we have for all $t \in (0, T_{\max})$ and almost all $x \in \mathcal{X}$,

$$p\,|u(t)|^{p-2} u(t)\, \partial_t^\alpha u(t) \geq \partial_t^\alpha \left(|u(t)|^p\right) = \frac{d}{dt}\Big(g_{1-\alpha} * \left[|u|^p - |u_0|^p\right]\Big)(t),$$

for any $p \in [2, \infty)$.

Proof By Theorem 3.2.6, any bounded mild solution is a strong solution on $[t, T]$, for any $0 < t \leq T < T_{\max}$, namely,

$$u \in C^{0,\kappa}\left([t, T]; L^\infty(\mathcal{X})\right),\ \partial_t^\alpha u \in C\left([t, T]; L^p(\mathcal{X})\right)$$

and

$$u \in C\left([t,T]; D\left(A_p\right)\right).$$

The statement is then a consequence of the proof of [18, Corollary 6.1] and a standard approximation argument (see [18, pg. 973]) that allows to replace the singular kernel $g_{1-\alpha}$ by its Yosida regularization $g_{1-\alpha,n} \in W^{1,1}[0,T]$. □

Remark 3.4.10 We note that alternatively the result of Proposition 3.4.9 is also a consequence of [20, Proposition 2.18] provided that $u \in C^1\left((0,T); L^p(\mathcal{X})\right) \cap C\left([0,T); L^p(\mathcal{X})\right)$. Such regularity is in fact readily available for the strong solution of (3.1.1) provided that some natural additional conditions are satisfied by the source $f = f(x,t,\xi)$ (see Sect. 3.3).

Let $T \in (0,\infty)$ be given and set $c_0 := \sup_{t\in(0,T)} \|c(\cdot,t)\|_{L^\infty(\mathcal{X})} > 0$. We thus assume that

(F6) $f(x,t,\xi)\xi \le c(x,t)\left(1+|\xi|^2\right)$, for all $\xi \in \mathbb{R}$, a.e. $(x,t) \in \mathcal{X}\times(0,\infty)$.

This additional assumption on f is quite natural in the classical theory for reaction–diffusion equations (see [1]). We note that **(F6)** is not an assumption about the growth of the nonlinearity as $|\xi| \to \infty$, but rather it is a coercivity condition since it allows for a large class of (dissipative) polynomial nonlinearities. In particular, the following example of

$$f(x,t,\xi) = -c_f|\xi|^n\xi + \sum_{i=0}^{n-1} c_{i+1}(x,t)|\xi|^i\xi + c_0(x,t),$$

for some $c_f > 0$, $c_i \in \mathbb{R}$ and $n \in \mathbb{N}$, is frequently encountered in applications dealing with the (large time) asymptotic behavior of solutions for (3.1.1) when $\alpha = 1$ (see, for instance, [11–13], and the references therein).

Theorem 3.4.11 (Global L^∞-Estimate) *Let the assumptions of Theorem 3.2.6 be satisfied. Suppose that the operator $A = A_2$ (as given by (2.2.7)) has $D(\mathcal{E}) = V \overset{d}{\hookrightarrow} L^2(\mathcal{X})$ and $V \hookrightarrow L^{2q}(\mathcal{X})$, for some $q := q_A > 1$. Under the assumption* ***(F6), any maximal mild (or quasi-mild) solution*** *of Problem (3.1.1), $\alpha \in (0,1)$, is* ***globally bounded*** *on $(0,\infty)$, namely, $T_{\max} = \infty$. In particular, the following estimate holds*

$$\sup_{t\in(0,T)} \|u(t)\|_{L^\infty(\mathcal{X})} \le C\left(\|u_0\|_{L^\infty(\mathcal{X})} + C_0E_{\alpha,1}\left(C_0T^\alpha\right) + T^\alpha E_{\alpha,\alpha}\left(C_0T^\alpha\right)\right), \tag{3.4.33}$$

for some $C = C(\alpha) > 0$ and $C_0 > 0$, both independent of t, T, u_0 and u. The constant $C = C(\alpha)$ is bounded as $\alpha \to 1^-$ and C_0 is independent of α.

Proof By the first hypothesis of the theorem, any mild solution is a strong solution on $[t,T]$, for all $0 < t \le T < T_{\max}$, and therefore, it satisfies the differential equation in (3.1.1), pointwise in time and for almost all $x \in \mathcal{X}$. Thus, we can multiply the equation by $p\,|u|^{p-2}u$, $p \in [2,\infty)$ and integrate the resulting identity over $\mathcal{X}$. Applying the inequality on the left of (2.2.11), namely,

$$4\frac{p-1}{p}\mathcal{E}\left(u|u|^{\frac{p}{2}-1}, u|u|^{\frac{p}{2}-1}\right) \le p\left(-A_p u, |u|^{p-2}u\right)_{L^2(\mathcal{X})},$$

we derive by virtue of Proposition 3.4.9, the following inequality:

$$\partial_t^{\alpha}(\|u\|^p_{L^p(\mathcal{X})}) + \frac{4(p-1)}{p}\mathcal{E}\left(u|u|^{\frac{p}{2}-1}, u|u|^{\frac{p}{2}-1}\right) \le pc_0\left(\mathfrak{m}(\mathcal{X}) + \|u\|^p_{L^p(\mathcal{X})}\right). \tag{3.4.34}$$

We note that (3.4.34) is the key in proving the desired estimate (3.4.33).

Step 1 (The iterative procedure). To this end, setting $p\,(=p_k) = 2^k$, $k \ge 1$, and

$$x_k(t) := \int_{\mathcal{X}} |u(t)|^{2^k}\,d\mathfrak{m}, \quad k \ge 1,$$

and having established (3.4.34), we obtain

$$\begin{aligned}\partial_t^{\alpha} x_k(t) &+ \frac{4p_k - 1}{p_k}\mathcal{E}\left(|u(t)|^{(p_k-2)/2}u(t), |u(t)|^{(p_k-2)/2}u(t)\right) \\ &\le Cp_k x_k(t) + Cp_k, \forall t \ge 0,\end{aligned} \tag{3.4.35}$$

for some $C = C(c_0, \mathfrak{m}(\mathcal{X})) > 0$. As usual, our goal is to derive a recursive inequality for x_k using (3.4.35). In order to do so, for $q > 1$ fixed such that $V \hookrightarrow L^{2q}(\Omega)$, we define

$$\overline{p}_k := \frac{p_k - p_{k-1}}{qp_k - p_{k-1}} = \frac{1}{2q-1} < 1, \; \overline{q}_k := 1 - \overline{p}_k = 2\frac{q-1}{2q-1}.$$

We aim to estimate the x_k-term on the right-hand side of (3.4.35) in terms of x_{k-1}. Next, the Hölder inequality, the Sobolev inequality $D(\mathcal{E}) = V \hookrightarrow L^{2q}(\Omega)$ together with the Poincaré inequality [10, Proposition 2.1], namely, for every $\eta \in (0,1)$, there exists $m > 0$ such that

$$\|v\|^2_{L^2(\mathcal{X})} \le \eta\mathcal{E}(v,v) + \eta^{-m}\|v\|^2_{L^1(\mathcal{X})} \tag{3.4.36}$$

yield

$$\begin{aligned} x_k = \int_{\mathcal{X}} |u|^{p_k}\,d\mathfrak{m} &\le \left(\int_{\mathcal{X}} |u|^{p_k q}\,d\mathfrak{m}\right)^{\overline{p}_k}\left(\int_{\mathcal{X}} |u|^{p_{k-1}}\,d\mathfrak{m}\right)^{\overline{q}_k} \\ &\le C\left[\mathcal{E}\left(|u|^{(p_k-2)/2}u, |u|^{(p_k-2)/2}u\right) + \left(\int_{\mathcal{X}} |u|^{p_{k-1}}\,d\mathfrak{m}\right)^2\right]^{\overline{s}_k}\left(\int_{\mathcal{X}} |u|^{p_{k-1}}\,d\mathfrak{m}\right)^{\overline{q}_k}. \end{aligned} \tag{3.4.37}$$

Here, $\overline{s}_k = \overline{p}_k q \equiv q/(2q-1) \in (0,1)$ (also note that $p_k/2 = p_{k-1}$). Here, note that we have also used that

$$\mathcal{E}\left(|u|^{p_{k-1}}, |u|^{p_{k-1}}\right) = \mathcal{E}\left(|u|^{(p_k-2)/2}u, |u|^{(p_k-2)/2}u\right), \text{ for } p_k = 2^k, \ k \geq 1,$$

which holds as a consequence of the fact that $|u|^{p_{k-1}} sgn(u) = |u|^{(p_k-2)/2}u$. Applying Young's inequality on the right-hand side of (3.4.37), we get for every $\eta > 0$,

$$p_k \int_{\mathcal{X}} |u|^{p_k} d\mathfrak{m} \leq \eta\mathcal{E}\left(|u|^{(p_k-2)/2}u, |u|^{(p_k-2)/2}u\right) + \eta\left(\int_{\mathcal{X}} |u|^{p_{k-1}} d\mathfrak{m}\right)^2 \quad (3.4.38)$$

$$+ Q_\eta(p_k)\left(\int_{\mathcal{X}} |u|^{p_{k-1}} d\mathfrak{m}\right)^2,$$

for some function $Q_\eta(\cdot) : \mathbb{R}_+ \to \mathbb{R}_+$ independent of k, owing to the fact that $z_k := \overline{q}_k/(1-\overline{s}_k) \equiv 2$ (indeed, $Q_\eta(y) = C_\eta y^{4/\overline{q}_k} = C_\eta y^{2(2q-1)/(q-1)}$, for some constant $C_\eta > 0$). Therefore, inserting (3.4.38) into the inequality (3.4.35), choosing a sufficiently small $\eta \in (0,1)$, independent of k, we obtain for $t \geq 0$,

$$\partial_t^\alpha x_k(t) + \mathcal{E}\left(|u(t)|^{(p_k-2)/2}u(t), |u(t)|^{(p_k-2)/2}u(t)\right) \quad (3.4.39)$$

$$\leq Q_\eta\left(2^k\right)(x_{k-1}(t))^2 + C2^k.$$

By application of (3.4.36) once more, we infer that

$$\eta\mathcal{E}\left(|u|^{(p_k-2)/2}u, |u|^{(p_k-2)/2}u\right) \geq \int_{\mathcal{X}} |u|^{p_k} d\mu - \eta^{-m}\left(\int_{\mathcal{X}} |u|^{p_{k-1}} d\mathfrak{m}\right)^2 \quad (3.4.40)$$

$$= x_k - \eta^{-m}(x_{k-1})^2,$$

for some $m > 0$ independent of u, k. We can now combine (3.4.40) with (3.4.39) to deduce

$$\partial_t^\alpha x_k(t) + x_k(t) \leq Q_\eta\left(2^k\right)(x_{k-1})^2 + C2^k, \ \forall t \geq 0. \quad (3.4.41)$$

The foregoing inequality together with the application of Lemma A.0.8, gives

$$x_k \leq C(\alpha)\max\left\{\int_{\mathcal{X}} |u_0|^{2^k} d\mathfrak{m}, Q_\eta\left(2^k\right)\sup_{t\geq 0}(x_{k-1}(t))^2 + C2^k\right\}, \quad (3.4.42)$$

for all $k \geq 1$. On the other hand, let us observe that there exists a positive constant $C_\infty = C_\infty(\|u_0\|_{L^\infty(\mathcal{X})}) \geq 1$, independent of k, such that $\|u_0\|_{L^{2^k}(\mathcal{X})} \leq C_\infty$. Taking the 2^k-th root on both sides of (3.4.42), and defining $M_k := \sup_{t\geq 0} \|u(t)\|_{L^{2^k}(\Omega)}$, we easily arrive at

$$M_k \leq C(\alpha) \max\left\{ C_\infty, \left(C_\eta \left(2^k\right)^{\frac{2(2q-1)}{q-1}} \right)^{\frac{1}{2^k}} M_{k-1} + C2^{k2^{-k}} \right\}, \text{ for all } k \geq 1. \tag{3.4.43}$$

We can now iterate in a standard way in (3.4.43) (see, for instance, [5, Lemma 9.3.1]), to finally obtain

$$\begin{aligned} \sup_{t\geq 0} \|u(t)\|_{L^\infty(\mathcal{X})} &\leq \lim_{k\to+\infty} M_k \\ &\leq C(\alpha) \max\left(C_\infty, \sup_{t\geq 0} \|u(t)\|_{L^2(\mathcal{X})} + 1 \right), \end{aligned} \tag{3.4.44}$$

for some $C(\alpha) > 0$ which is bounded as $\alpha \to 1$ but is independent of $t \geq 0$.

Step 2 (The $L_t^\infty L_x^2$-estimate). In this final step, we derive the required $L^\infty((0,T); L^2(\mathcal{X}))$-bound, for any $T > 0$; this combined together with (3.4.44) gives the desired inequality (3.4.33). The latter also shows in particular that one can take $T_{\max} = \infty$. Taking $k = 1$ into (3.4.41), we deduce

$$\partial_t^\alpha (\|u(t)\|_{L^2(\mathcal{X})}^2) \leq C_0 \left(\|u(t)\|_{L^2(\mathcal{X})}^2 + 1 \right), \ \forall t \geq 0,$$

for some $C_0 > 0$ independent of time and α. Consider now the linear problem, given by $\partial_t^\alpha z = C_0(z+1)$, $z(0) = \|u_0\|_{L^2(\mathcal{X})}^2 \geq 0$, and observe that its unique solution (see the proof of Lemma A.0.8 and (A.0.19)) is given by

$$z(t) = E_{\alpha,1}(C_0 t^\alpha) z(0) + C_0^{1-1/\alpha} t^\alpha E_{\alpha,\alpha+1}(C_0 t^\alpha), \ t \geq 0.$$

The comparison principle in Lemma A.0.7 then gives

$$\|u(t)\|_{L^2(\mathcal{X})}^2 \leq E_{\alpha,1}(C_0 t^\alpha) \|u_0\|_{L^2(\mathcal{X})}^2 + C_0^{1-1/\alpha} t^\alpha E_{\alpha,\alpha+1}(C_0 t^\alpha),$$

for any $t \geq 0$. Hence, the smoothing property (3.4.44) immediately entails the assertion of the theorem since $E_{\alpha,\alpha+1}(\tau) \leq (1/\alpha) E_{\alpha,\alpha}(\tau)$ for $\tau \geq 0$. □

We can improve the previous result so that the bound is also uniform with respect to the final time $T > 0$.

Corollary 3.4.12 *Let the assumptions of Theorem 3.4.11 be satisfied and assume that $c_0 < \lambda_1$ where $\lambda_1 = \inf \sigma(-A) > 0$ is the first eigenvalue of $-A$. Then the mild (or quasi-mild) solution is bounded in $L^\infty((0,T); L^\infty(\mathcal{X}))$, also uniformly with respect to $T > 0$.*

Proof By (3.4.34) with $p = 2$, we have

$$\partial_t^\alpha \left(\|u\|_{L^2(\mathcal{X})}^2\right) + 2\mathcal{E}(u,u) \leq 2c_0 \left(\mathrm{m}(\mathcal{X}) + \|u\|_{L^2(\mathcal{X})}^2\right)$$

which yields, for $d_0 := 2(\lambda_1 - c_0) > 0$,

$$\partial_t^\alpha \left(\|u\|_{L^2(\mathcal{X})}^2\right) + d_0 \|u\|_{L^2(\mathcal{X})}^2 \leq 2c_0 \mathrm{m}(\mathcal{X}), \text{ for } t \in (0,T).$$

The application of Lemma A.0.8 then gives

$$\|u(t)\|_{L^2(\mathcal{X})}^2 \leq 2C(\alpha) \max \left\{ \|u_0\|_{L^2(\mathcal{X})}^2, \frac{2c_0 \mathrm{m}(\mathcal{X})}{d_0^{1/\alpha}} \right\}.$$

Inserting this once again into (3.4.44), we immediately get the desired claim. □

Corollary 3.4.13 *The conclusions of Theorem 3.4.11 as well as of Corollary 3.4.12 are valid for the entire class of diffusion operators A, as given in Sect. 2.3.*

Remark 3.4.14 Note that in the limit of $\alpha \to 1$, we recover the global estimate stated in Corollary 3.4.13 for the classical diffusion problem (1.0.1) with $\alpha = 1$. This result can be seen as a generalization of a number of similar results proven recently in [11–15].

3.5 Limiting Behavior as $\alpha \to 1$

We conclude this section with a convergence result as $\alpha \to 1^-$, which in light of the previous two results, is now possible due to the uniform $L_t^\infty L_x^\infty$-bound. For the sake of convenience and to simplify any further technicalities, we also take $c \in L_{\infty,\infty}$ in assumptions (**F3**)–(**F4**) (i.e., $q_1 = q_2 = \infty$), and elsewhere in the assumptions of the corresponding theorems. Let $T > 0$ be given, but otherwise arbitrary.

Theorem 3.5.1 *Let the assumptions of Theorem 3.4.11 be satisfied and let $u_0 \in \mathcal{L}^\infty(\mathcal{X})$. Let v be the bounded mild solution for*

$$\partial_t v = Av + f(x,t,v), \ (x,t) \in \mathcal{X} \times (0,T], \ v_{|t=0} = u_0$$

and let $u = u_\alpha$ be the corresponding bounded mild solution of (3.1.1) when $\alpha \in (0,1)$. Then the following convergence holds:

$$\lim_{\alpha \to 1^-} \sup_{t \in [T_0, T]} \|u_\alpha(t) - v(t)\|_{L^\infty(\mathcal{X})} = 0, \tag{3.5.1}$$

for any $0 < T_0 \le T < \infty$.

Proof We first recall that

$$u_\alpha(t) = S_\alpha(t) u_0 + \int_0^t P_\alpha(t-s) f(s, u_\alpha(s))\, ds$$

and

$$v(t) = S(t) u_0 + \int_0^t S(t-s) f(s, v(s))\, ds.$$

By [6, Theorem 2.42], we have $S_\alpha(t) = E_{\alpha,1}(t^\alpha A)$ and $P_\alpha(t) = t^{\alpha-1} E_{\alpha,\alpha}(t^\alpha A)$, $\alpha \in (0,1)$, where both operators $E_{\alpha,1}(t^\alpha A)$ and $E_{\alpha,\alpha}(t^\alpha A)$ have a Cauchy integral representation over a proper Hankel path, see [6, Theorem 2.41]. Application of [6, Lemma 3.12] then yields, for all $u_0 \in \mathcal{L}^\infty(\mathcal{X})$, $S_\alpha(t) u_0 \to S(t) u_0$ and $P_\alpha(t) u_0 \to S(t) u_0$, as $\alpha \to 1^-$; these convergences are also uniform on bounded subsets of $\mathcal{L}^\infty(\mathcal{X})$ and on intervals $[T_0, T]$, for any $T_0 > 0$. Since $u_\alpha, v \in L^\infty((0,T); L^\infty(\mathcal{X}))$ uniformly with respect to $\alpha \to 1^-$, we obtain on account of the Lebesgue dominated convergence theorem, that

$$\lim_{\alpha \to 1^-} \left\| \int_0^t (P_\alpha(t-s) - S(t-s)) f(s, v(s))\, ds \right\|_{L^\infty(\mathcal{X})} = 0. \tag{3.5.2}$$

Taking the difference between u_α and v, we then see that

$$\begin{aligned} &\|u_\alpha(t) - v(t)\|_{L^\infty(\mathcal{X})} \\ &\le A_\alpha(t) + \left\| \int_0^t P_\alpha(t-s)(f(s, u_\alpha(s)) - f(s, v(s)))\, ds \right\|_{L^\infty(\mathcal{X})}, \end{aligned} \tag{3.5.3}$$

where

$$\begin{aligned} A_\alpha(t) &:= \|S_\alpha(t) u_0 - S(t) u_0\|_{L^\infty(\mathcal{X})} \\ &+ \left\| \int_0^t (P_\alpha(t-s) - S(t-s)) f(s, v(s))\, ds \right\|_{L^\infty(\mathcal{X})}. \end{aligned} \tag{3.5.4}$$

Note that $A_\alpha(t) \to 0$ as $\alpha \to 1^-$ in light of (3.5.2) and the fact that $S_\alpha(t) u_0 \to S(t) u_0$. Since f is also locally Lipschitz owing to (**F4**), we further derive from (3.5.3) that

$$\|u_\alpha(t) - v(t)\|_{L^\infty(\mathcal{X})} \le A_\alpha(t) + \frac{Q(M)}{\Gamma(\alpha)} \int_0^t (t-s)^{\alpha-1} \|u_\alpha(s) - v(s)\|_{L^\infty(\mathcal{X})} ds, \tag{3.5.5}$$

for all $t \in (0, T)$, where $Q(M) > 0$ is independent of $\alpha \to 1^-$. Here, we have set $M := \sup_{t\in(0,T)} \max\{\|u_\alpha(t)\|_{L^\infty(\mathcal{X})}, \|v(t)\|_{L^\infty(\mathcal{X})}\}$. We infer from (3.5.5) and the application of the Gronwall lemma (see Lemma A.0.9), that

$$\begin{aligned} &\|u_\alpha(t) - v(t)\|_{L^\infty(\mathcal{X})} \\ &\le A_\alpha(t) + Q(M) \int_0^t (t-s)^{\alpha-1} E_{\alpha,\alpha}\left(Q(M)(t-s)^\alpha\right) A_\alpha(s) ds. \end{aligned} \tag{3.5.6}$$

The right-hand side of (3.5.6) converges to zero as $\alpha \to 1^-$, by virtue of (3.5.4) and the Lebesgue dominated convergence theorem (for the last summand). The proof is finished. □

3.6 Nonnegativity of Mild Solutions

We show that each mild solution constructed in Sect. 3.1 is nonnegative on its maximal interval of existence. We shall take advantage of their strong regularity proven in Sect. 3.2.

Theorem 3.6.1 *Let the assumptions of Theorem 3.1.10 be satisfied and assume that*

$$f(x, t, 0) \ge 0, \textit{ for a.e. } (x, t) \in \mathcal{X} \times (0, \infty).$$

If $u_0 \ge 0$ *then the mild solution of problem* (3.1.1) *satisfies* $u \ge 0$ *on* $(0, T_{\max})$.

Proof

Step 1. We first prove the claim for a (quasi) mild solution on $[0, T]$, $T < T_{\max}$, associated with an initial datum $u_0 \in L^\infty(\mathcal{X})$. We need to mollify both the nonlinearinity f and the function $c = c(x, t)$. This can be done by following a similar procedure as in the construction exploited in the proof of [22, Lemma 10, pg. 43]. Thus we may infer the existence of families $\{f_k\} \subset L_{\infty,\infty,T}$ and $\{c_k\} \subset L_{\infty,\infty,T}$, such that:

- For a.e. $(x, t) \in \mathcal{X} \times (0, \infty)$ and $\xi, \eta \in \mathbb{R}$,

$$\begin{cases} |f_k(x, t, \xi)| \le c_k(x, t) Q(|\xi|), \\ |f_k(x, t, \xi) - f_k(x, t, \eta)| \le c_k(x, t) Q(|\xi| + |\eta|) |\xi - \eta|, \\ f_k(x, t, 0) \ge 0. \end{cases} \tag{3.6.1}$$

- The functions $f_k : \xi \in [-U, U] \mapsto f_k(\cdot, \cdot, \xi) \in L_{q_1,q_2,T}$ are uniformly bounded and equicontinuous for all $k \in \mathbb{N}$,

$$\lim_{k\to\infty} \sup_{\xi\in[-U,U]} \|(f - f_k)(\cdot, \cdot, \xi)\|_{q_1,q_2,T} = 0. \tag{3.6.2}$$

- The functions c_k satisfy $\sup_{k\in\mathbb{N}} \|c_k\|_{q_1,q_2,T} \leq C \|c\|_{q_1,q_2,T}$, for some $C > 0$ independent of k, t and T.

These properties then imply that each bounded (quasi) mild solution u_k associated with the nonlinearity f_k, given by

$$u_k(t) = S_\alpha(t) u_0 + \int_0^t P_\alpha(t - \tau) f_k(\tau, u_k(\tau)) d\tau, \tag{3.6.3}$$

satisfies $\sup_{k\in\mathbb{N}} |||u_k|||_{\infty,0,T} \leq U$ (see the proof of Lemma 3.1.8). Moreover, u_k is a strong solution on $(0, T_{\max})$ in the sense of Definition 3.2.1 (namely, (b)–(d) are satisfied for a sufficiently large $p \in (\beta_A, \infty) \cap [2, \infty))$ by Theorem 3.2.6.

Let $H \in C^1(\mathbb{R})$ be given by $H(y) = (1/2) y^2$ for $y \in (-\infty, 0)$ and $H(y) = 0$ for $y \geq 0$, and notice that H is convex and $H'(y) = \min\{y, 0\} \leq 0$ for all $y \in \mathbb{R}$. Since u_k is a strong solution, it pointwise satisfies the equation

$$\partial_t^\alpha u_k = A u_k + f_k(\cdot, \cdot, u_k). \tag{3.6.4}$$

Multiplying (3.6.4) by $H'(u_k)$ and integrating over $\mathcal{X}$, we immediately find

$$\int_{\mathcal{X}} H'(u_k) \partial_t^\alpha u_k dm + \left(-A u_k, H'(u_k)\right)_{L^2(\mathcal{X})} \tag{3.6.5}$$
$$= \int_{\mathcal{X}} H'(u_k) f_k(\cdot, \cdot, u_k) dm.$$

Set now $\psi(t) = \int_{\mathcal{X}} H(u_k(t)) dm$ and observe that

$$\partial_t \psi(t) = \int_{\mathcal{X}} H'(u_k(t)) \partial_t u_k(t) dm \tag{3.6.6}$$

when $\alpha = 1$ (see [26, (1.100), pg. 52]), while in the case $\alpha \in (0, 1)$, by Proposition 3.4.9 it holds

$$\int_{\mathcal{X}} H'(u_k(t)) \partial_t^\alpha u_k(t) dm \geq \partial_t^\alpha \psi(t). \tag{3.6.7}$$

On the other hand, since $S(t)$ is positive by assumption (**HA**), application of Remark 2.2.6-(i) gives

$$\left(-Au_k, H'(u_k)\right)_{L^2(\mathcal{X})} = \mathcal{E}\left(u_k, H'(u_k)\right) \geq 0 \tag{3.6.8}$$

since $H'(u_k) = -u_k^-$. For the term on the right-hand side of (3.6.5), we argue as follows:

$$\int_{\mathcal{X}} H'(u_k) f_k(x,t,u_k)\,dm \tag{3.6.9}$$

$$= \int_{\mathcal{X}} H'(u_k)(f_k(x,t,u_k) - f_k(x,t,0))\,dm + \int_{\mathcal{X}} H'(u_k) f_k(x,t,0)\,dm$$

$$\overset{(3.6.1)}{\leq} \int_{\mathcal{X}} H'(u_k)(f_k(x,t,u_k) - f_k(x,t,0))\,dm$$

$$\overset{(3.6.1)}{\leq} \int_{\mathcal{X}} \left|H'(u_k)\right| |u_k| \, \|c_k\|_{\infty,\infty,T}\, Q(|u_k|)\,dm$$

$$\leq 2\,\|H(u)\|_{L^1(\mathcal{X})}\, Q(U)\,\|c_k\|_{\infty,\infty,T}\,.$$

Putting the estimates (3.6.6)–(3.6.9) together into Eq. (3.6.5), we obtain

$$\partial_t^{\alpha}\psi(t) \leq \chi\psi(t)\,,\ \chi := 2Q(U)\,\|c_k\|_{\infty,\infty,T}\,,$$

for all $t \in [0,T]$, $T < T_{\max}$. Application of the comparison principle (see Lemma A.0.7) in the case $\alpha \in (0,1)$, together with the Gronwall inequality when $\alpha = 1$, yields[3]

$$\psi(t) \leq \psi(0)\, E_{\alpha,1}\left(\chi t^{\alpha}\right),$$

for all $t \in [0,T]$ and $\alpha \in (0,1]$. Hence, since $u_k(0) = u_0 \geq 0$, $\psi(0) = 0$ implies that $\psi(t) = 0$, namely, $u_k(t) \geq 0$ for all $t \in [0,T]$.

Step 2 (The limit procedure). We take the limit as $k \to \infty$ in the integral solution (3.6.3) by arguing exactly as in the proof of Lemma 3.1.8. We observe that

$$u_k - u = \int_0^t P_\alpha(t-\tau)(f(\tau,u(\tau)) - f(\tau,u_k(\tau)))\,d\tau$$
$$+ \int_0^t P_\alpha(t-\tau)(f(\tau,u_k(\tau)) - f_k(\tau,u_k(\tau)))\,d\tau.$$

[3] The Mittag-Leffler function $E_{1,1}(x) = e^x$.

Next, as in the proof of Lemma 3.1.8, choose a sufficiently small time $T_* \in (0, 1]$ such that

$$CT_*^{\varepsilon} \|c\|_{q_1,q_2} Q(2U) \leq \frac{1}{2}. \tag{3.6.10}$$

Analogously to arguments leading up to the estimate (3.1.26), by virtue of (3.6.1) and the assumptions on f, we arrive at

$$\begin{aligned} |||u_k - u|||_{\infty,0,T_*} \leq &CT_*^{\varepsilon} \|c\|_{q_1,q_2} Q(2U) |||u_k - u|||_{\infty,0,T_*} \\ &+C \sup_{u \in [-U,U]} \|(f - f_k)(\cdot, \cdot, u)\|_{q_1,q_2,T_*}, \end{aligned} \tag{3.6.11}$$

for some $C > 0$ independent of k, t and T_*. In particular, (3.6.10) and (3.6.2) yield

$$\lim_{k \to \infty} |||u_k - u|||_{\infty,0,T_*} = 0.$$

Now $u_k \geq 0$ on $[0, T_*]$ implies that $u \geq 0$ on $[0, T_*]$, and a standard continuation argument allows us to extend this property on $[0, T]$, with $T < T_{\max}$.

Step 3 (The extension). We extend the above argument to a local mild solution associated with an initial datum $u_0 \in L^{p_0}(\mathcal{X})$, with $p_0 \in [1, \infty)$, satisfying all the assumptions of Theorem 3.1.4-(a). We choose a sequence $u_{0n} \in \mathcal{L}^{\infty}(\mathcal{X})$ such that $\|u_{0n} - u_0\|_{L^{p_0}(\mathcal{X})} \to 0$ as $n \to \infty$, with $u_{0n} \geq 0$ and $\|u_{0n}\|_{L^{p_0}(\mathcal{X})} \leq \|u_0\|_{L^{p_0}(\mathcal{X})}$, for all $n \in \mathbb{N}$. We then infer from Lemma 3.1.5 the existence $T \in (0, 1]$ and $U < \infty$ and a (quasi) mild bounded solution $u_n \in E_{p,\delta,T}$ for the initial datum u_{0n}, on the interval $[0, T]$, such that

$$\sup_{n \in \mathbb{N}} |||u_n|||_{p,\delta,T} \leq U. \tag{3.6.12}$$

Then similarly to estimates (3.1.10)–(3.1.12), we derive that

$$\begin{aligned} |||u_n - u|||_{p,\delta,T} \leq &C \|u_{0n} - u_0\|_{L^{p_0}(\mathcal{X})} \\ &+ CT^{\varepsilon} \|c\|_{q_1,q_2} |||1 + |u_n| + |u|||_{p,\delta,T}^{\gamma-1} |||u_n - u|||_{p,\delta,T} \end{aligned}$$

for a time $T \leq 1$ that satisfies

$$CT^{\varepsilon} (1 + 2U)^{\gamma-1} \leq \frac{1}{2}.$$

Therefore, (3.6.12) yields

$$|||u_n - u|||_{p,\delta,T} \leq C \|u_{0n} - u_0\|_{L^{p_0}(\mathcal{X})}.$$

Since $u_n \geq 0$ for all $n \in \mathbb{N}$ by Step 2, passing to the limit as $n \to \infty$ in the foregoing inequality gives that $u \geq 0$ on $[0, T]$. In the case when $0 \leq u_0 \in L^{p_0}(\mathcal{X}) \setminus \{0\}$, with $p_0 \in (1, \infty)$ satisfying the assumptions of Theorem 3.1.4-(c), we can choose a sequence $u_{0n} \in L^\infty(\mathcal{X})$ such that $\Pi := \{u_0, u_{0n} : n \in \mathbb{N}\} \subset L^{p_0}(\mathcal{X}) \setminus \{0\}$. Then one argues instead in the space $E_{w,p,\delta,T}$ by means of the proof of Lemma 3.1.6. Since the details remain the same as above but for some minor modifications, we may omit them completely and thus conclude the proof of the theorem. □

3.7 An Application: The Fractional Fisher-KPP Equation

As one straight-forward application of our abstract results in the previous sections, we may consider the Fisher-KPP equation with fractional-in-time derivative

$$\partial_t^\alpha u = Au + f(u) \text{ in } \Omega \times (0, \infty), \ u_{|t=0} = u_0 \text{ in } \Omega, \tag{3.7.1}$$

and a diffusion operator A that satisfies the assumption (**HA**). The nonlinearity $f \in C[0,1]$ is such that

$$f > 0 \text{ on } (0,1) \text{ with } f(0) = f(1) = 0. \tag{3.7.2}$$

The basic instructive example is $f(u) = u(1-u)$. Here, Ω denotes a bounded domain with sufficiently smooth boundary. An important example is $A = \Delta$ (see Example 2.3.5) but also the fractional Laplacian $(-\Delta)_\Omega^s$, we refer to Example 2.3.6 in Sect. 2.3. For the latter example, in the case $\alpha = 1$, the existence of a unique bounded mild solution was recently obtained in [4], where the large time behavior of front solutions has also been investigated. We mention that the problem (3.7.1)–(3.7.2) has been also considered in [19] in the case where A is the discrete fractional Laplace operator. We point out that the occurrence of an exponent $\alpha \in (0,1)$ in (3.7.1), other than the classical one $\alpha = 1$, is a consequence of physical phenomena in reactive systems (such as, plasma, flames and chaotic phase-transitions) whose diffusive behavior is anomalous (see [21]). We refer the reader to Appendix C for a complete description of models that contain fractional kinetics. In the context of phase-transitions for binary materials, Eq. (3.7.1) for $\alpha = 1$ is usually referred as the Allen-Cahn equation, where u denotes the atom concentration of one of the material components (see, for instance, [13]).

Our first result is on the existence of global strong solutions for (3.7.1). We conveniently extend f by continuity to the whole of $\mathbb{R}$ such that $f < 0$ on $(-\infty, 0) \cup (1, \infty)$. We recall once again that A_p stands for the generator of the associated semigroup $S(t)$ on $L^p(\Omega)$. For $p \geq 2$, such generator possesses even an explicit characterization (2.2.10).

Theorem 3.7.1 *Let* (3.7.2) *be satisfied and assume* $0 \leq u_0 \leq 1$, $u_0 \in D(A_p)$ *for* $p \in (\beta_A, \infty) \cap (1, \infty)$. *Then problem* (3.7.1), $\alpha \in (0,1]$, *admits a unique globally-defined strong solution in the sense of Definition 3.2.1 such that* $0 \leq u \leq 1$.

Proof Application of Corollary 3.2.3 and Theorem 3.6.1 yields the existence of a strong solution u on a maximal interval $(0, T_{\max})$ and $u \geq 0$ on $(0, T_{\max})$. The time $T_{\max} > 0$ is such that either $T_{\max} = \infty$ or $T_{\max} < \infty$ and $\lim_{t \to T_{\max}} \|u(t)\|_{L^\infty(\Omega)} = \infty$. In order to show that $T_{\max} = \infty$, it then suffices to show that $u \leq 1$ for all $t > 0$. To this end, define $H(y) = \frac{1}{2}(y-1)^2$, for $y > 1$, and $H(y) = 0$, for $y \in [0, 1]$ and observe that $H \in C^2[0, \infty)$ is convex. As in the proof of Theorem 3.6.1 (see Step 1, (3.6.7)–(3.6.8)), setting $\psi(t) = \int_\Omega H(u(t))\,dx$ and then testing (3.7.1) with $H'(u)$, gives

$$\partial_t^\alpha H(u) \leq \int_{\{x \in \Omega : u > 1\}} f(u)(u-1)\,dx \leq 0.$$

The comparison principle (Lemma A.0.7) yields $H(u(t)) \leq H(u_0) = 0$ and therefore, $u \leq 1$ pointwise in time. This concludes the proof. □

For rough initial datum, we deduce the following well-posedness result.

Theorem 3.7.2 *Let $u_0 \in L^\infty(\Omega)$ such that $u_0 \in [0, 1]$. Then Problem (3.7.1) has a unique global and bounded quasi-mild solution in the sense of Definition 3.1.3 such that $0 \leq u \leq 1$. This solution also satisfies*

$$\lim_{t \to 0} \|u(t) - u_0\|_{L^p(\Omega)} = 0, \ \forall p \in [1, \infty),$$

such that $p = \infty$, if we additionally assume that $u_0 \in \mathcal{L}^\infty(\Omega)$. Furthermore, this mild solution is also a strong solution on $[T_0, \infty)$, for every $T_0 > 0$.

Proof The proof follows from Theorem 3.7.1 and an approximation procedure similar to Step 2 of Theorem 3.6.1. The last statement of the theorem is a consequence of Theorem 3.2.6. □

We can conclude with a basic comparison between solutions of (3.7.1) in the cases when $\alpha \in (0, 1)$ and $\alpha = 1$, respectively. Indeed, by Theorem 3.5.1 it follows

Corollary 3.7.3 *Every bounded mild solution $u = u_\alpha$ of (3.7.1) converges strongly as $\alpha \to 1$ in the sense of (3.5.1) on the interval $(0, \infty)$, to a bounded mild solution of problem (3.7.1) with $\alpha = 1$.*

References

1. N.D. Alikakos, L^p bounds of solutions of reaction-diffusion equations. Commun. Partial Differ. Equ. **4**(8), 827–868 (1979)
2. U. Biccari, M. Warma, E. Zuazua, Addendum: Local elliptic regularity for the Dirichlet fractional Laplacian [MR3641649]. Adv. Nonlinear Stud. **17**(4), 837–839 (2017)
3. U. Biccari, M. Warma, E. Zuazua, Local elliptic regularity for the Dirichlet fractional Laplacian. Adv. Nonlinear Stud. **17**(2), 387–409 (2017)

4. X. Cabré, J.-M. Roquejoffre, The influence of fractional diffusion in Fisher-KPP equations. Commun. Math. Phys. **320**(3), 679–722 (2013)
5. J.W. Cholewa, T. Dlotko, *Global Attractors in Abstract Parabolic Problems*. London Mathematical Society Lecture Note Series, vol. 278 (Cambridge University Press, Cambridge, 2000)
6. P.M. de Carvalho Neto, Fractional Differential Equations: A Novel Study of Local and Global Solutions in Banach Spaces. PhD thesis, Universidade de São Paulo, 2013
7. J. Diestel, J.J. Uhl Jr., *Vector Measures* (American Mathematical Society, Providence, RI, 1977). With a foreword by B. J. Pettis, Mathematical Surveys, No. 15
8. K. Diethelm, *The Analysis of Fractional Differential Equations*. Lecture Notes in Mathematics, vol. 2004 (Springer, Berlin, 2010)
9. H.O. Fattorini, *Infinite-Dimensional Optimization and Control Theory*. Encyclopedia of Mathematics and Its Applications, vol. 62 (Cambridge University Press, Cambridge, 1999)
10. C.G. Gal, Doubly nonlocal Cahn-Hilliard equations. Ann. Inst. H. Poincaré Anal. Non Linéaire **35**(2), 357–392 (2018)
11. C.G. Gal, M. Warma, Elliptic and parabolic equations with fractional diffusion and dynamic boundary conditions. Evol. Equ. Control Theory **5**(1), 61–103 (2016)
12. C.G. Gal, M. Warma, Long-term behavior of reaction-diffusion equations with nonlocal boundary conditions on rough domains. Z. Angew. Math. Phys. **67**(4), 42 (2016). Art. 83
13. C.G. Gal, M. Warma, Reaction-diffusion equations with fractional diffusion on non-smooth domains with various boundary conditions. Discrete Contin. Dyn. Syst. **36**(3), 1279–1319 (2016)
14. C.G. Gal, M. Warma, Transmission problems with nonlocal boundary conditions and rough dynamic interfaces. Nonlinearity **29**(1), 161–197 (2016)
15. C.G. Gal, M. Warma, Nonlocal transmission problems with fractional diffusion and boundary conditions on non-smooth interfaces. Commun. Partial Differ. Equ. **42**(4), 579–625 (2017)
16. G. Grubb, Regularity of spectral fractional Dirichlet and Neumann problems. Math. Nachr. **289**(7), 831–844 (2016)
17. D. Henry, *Geometric Theory of Semilinear Parabolic Equations*. Lecture Notes in Mathematics, vol. 840 (Springer, Berlin/New York, 1981)
18. J. Kemppainen, J. Siljander, V. Vergara, R. Zacher, Decay estimates for time-fractional and other non-local in time subdiffusion equations in $\mathbb{R}^d$. Math. Ann. **366**(3–4), 941–979 (2016)
19. V. Keyantuo, C. Lizama, M. Warma, Lattice dynamical systems associated with a fractional Laplacian. Numer. Funct. Anal. Optim. **40**(11), 1315–1343 (2019)
20. J.-G. Li, L. Liu, Some compactness criteria for weak solutions of time fractional PDEs. SIAM J. Math. Anal. **50**, 3693–3995 (2018)
21. R. Mancinelli, D. Vergni, A. Vulpiani, Front propagation in reactive systems with anomalous diffusion. Phys. D **185**(3–4), 175–195 (2003)
22. F. Rothe, *Global Solutions of Reaction-Diffusion Systems*. Lecture Notes in Mathematics, vol. 1072 (Springer, Berlin, 1984)
23. H.-J. Schmeisser, H. Triebel (eds.) *Function Spaces, Differential Operators and Nonlinear Analysis*. Teubner-Texte zur Mathematik [Teubner Texts in Mathematics], vol. 133 (B. G. Teubner Verlagsgesellschaft mbH, Stuttgart, 1993). Papers from the International Conference held in Friedrichroda, September 20–26, 1992
24. V. Vergara, R. Zacher, A priori bounds for degenerate and singular evolutionary partial integro-differential equations. Nonlinear Anal. **73**(11), 3572–3585 (2010)
25. V. Vergara, R. Zacher, Optimal decay estimates for time-fractional and other nonlocal subdiffusion equations via energy methods. SIAM J. Math. Anal. **47**(1), 210–239 (2015)
26. A. Yagi, *Abstract Parabolic Evolution Equations and Their Applications*. Springer Monographs in Mathematics (Springer, Berlin, 2010)

Chapter 4
Systems of Fractional Kinetic Equations

In this chapter, we consider some general classes of reaction–diffusion systems that contain some fractional kinetics occurring in applications (cf. Appendix C below), and then investigate their local and global existence of solutions in detail. In a preliminary step, we derive results that allow for the existence of sufficiently smooth solutions which are needed in order to rigorously justify other precise and explicit calculations (namely, maximum principles, energy estimates and comparison arguments) which will be performed on more specific models in the sequel. It turns out that the techniques employed for the scalar equation (3.1.1) in the previous chapter will prove quite useful in the analysis.

4.1 Nonlinear Fractional Reaction–Diffusion Systems

Let $m \in \mathbb{N}$ and $u = (u_1, \ldots, u_m) \in \mathbb{R}^m$ where each u_i $(i = 1, \ldots, m)$ is a measurable physical quantity. Let $d_i = 0$ for $i = 1, \ldots, r$ and $d_i > 0$ for $i = r + 1, \ldots, m$. We allow the case $r = 0$ to occur so that all $d_i > 0$ for $i = 1, \ldots, m$ in some cases. Next, let $D = \operatorname{diag}(d_1, \ldots, d_m)$ be the diagonal matrix of diffusion coefficients and assume that $u_0 = (u_{01}, \ldots, u_{0m})(x) \in \mathbb{R}^m$, for $x \in \mathcal{X}$, models the initial data. Let $f = (f_1, \ldots, f_m)(x, t, u_1, \ldots, u_m)$ with

$$f : (x, t, u) \in \mathcal{X} \times [0, \infty) \times \mathbb{R}^m \to f(x, t, u) \in \mathbb{R}^m,$$

be a nonlinear function that models possible interactions between the various quantities u_i $(i = 1, \ldots, m)$. Finally, consider a family of closed operators $(A_i)_{i=1}^m$ that satisfies the assumptions of Propositions 2.2.1, 2.2.2. Namely, we assume that each A_i satisfies assumption (**HA**) with a possible different value $\beta_{A_i} > 0$ for $i = 1, \ldots, m$. In particular, let $(S_i(t))_{i=1}^m$ be the corresponding family of analytic semigroups associated with A_i, each S_i $(i = 1, \ldots, m)$ can be extended to a

C. G. Gal, M. Warma, *Fractional-in-Time Semilinear Parabolic Equations and Applications*, Mathématiques et Applications 84,
https://doi.org/10.1007/978-3-030-45043-4_4

contraction semigroup $S_{i,p_i}(t)$ on $L^{p_i}(\mathcal{X})$, $p_i \in [1, \infty]$, whose generator is A_{i,p_i} such that $A_{i,2} = A_i$. For any $\alpha_i \in (0, 1)$, the corresponding operators (2.1.9) associated with these semigroups S_i can also be defined analogously, to give the families of operators $\left(S_{\alpha_i}(t)\right)_{i=1}^{m}$ and $\left(P_{\alpha_i}(t)\right)_{i=1}^{m}$, satisfying the ultracontractivity estimates of (2.2.12), after setting $\beta_A := \beta_{A_i}$, $\alpha := \alpha_i$. Finally, we set the diagonal (matrix) operator $A = \text{diag}(A_1, \ldots, A_m)$ and introduce the following notion of Caputo-fractional derivative

$$\partial_t^{\alpha} u = \left(\partial_t^{\alpha_1} u_1, \ldots, \partial_t^{\alpha_m} u_m\right) \in \mathbb{R}^m,$$

where each $\partial_t^{\alpha_i} u_i \in \mathbb{R}$ is understood in the sense of Definition 2.1.1 for $\alpha_i \in (0, 1)$. As usual, when $\alpha_i = 1$, $\partial_t^1 = \partial_t = d/dt$.

Our problem is to look for solutions $u = (u_1, \ldots, u_m)(x, t) \in \mathbb{R}^m$ of the following system

$$\partial_t^{\alpha} u = DAu + f(x, t, u), \quad (x, t) \in \mathcal{X} \times (0, \infty), \tag{4.1.1}$$

subject to the initial condition

$$u_{|t=0} = u_0 \text{ in } \mathcal{X}. \tag{4.1.2}$$

Note that components which do not diffuse as well as different kinds of "diffusion" operators for the diffusing components may occur in (4.1.1)–(4.1.2). Our goal is to construct bounded mild solutions for this initial-value problem and then turn to strong solutions. To this end, our assumptions on the nonlinearity f, from Sect. 3, are adapted to our new case, as follows.

(SF1) $f(x, t, \cdot) : \mathbb{R}^m \to \mathbb{R}^m$ is a measurable function for all $(x, t) \in \mathcal{X} \times (0, \infty)$ such that, for every bounded set $B \subset \overline{\mathcal{X}} \times [0, \infty) \times \mathbb{R}^m$, there exists a constant $L = L(B) > 0$ such that

$$|f(x, t, \xi)| \leq L(B), \quad \text{for all } (x, t, \xi) \in B$$

and

$$|f(x, t, \xi) - f(x, t, \eta)| \leq L(B)\, |\xi - \eta|, \quad \text{for all } (x, t, \xi), (x, t, \eta) \in B.$$

(SF2) For every bounded $B \subset \overline{\mathcal{X}} \times [0, \infty) \times \mathbb{R}^m$, there exists a constant $L(B) > 0$ such that, for all $(x, t, \xi), (x, s, \eta) \in B$,

$$|f(x, t, \xi) - f(x, s, \eta)| \leq L(B) \left(|t - s|^{\gamma} + |\xi - \eta|\right),$$

for some $\gamma > 0$.

As in Chap. 3, the assumption (**SF2**) will only be needed when the nonlinear source f is also time dependent; when $f = f(x, \xi)$ this condition is no longer necessary for the existence of strong solutions, we refer the reader to the proof of Corollary 3.2.3 (and Theorem 4.1.3 below).

First, we consider the initial-value problem (4.1.1)–(4.1.2) with $f \equiv 0$ and let $u_0 \in L^\infty(\mathcal{X}, \mathbb{R}^m)$. The solution $u = u(x,t)$ of the linear system for bounded initial datum u_0 defines a formal solution operator in the space $L^\infty(\mathcal{X}, \mathbb{R}^m)$ by setting

$$u(x,t) = (\mathbb{S}_\alpha(t) u_0)(x) = \left(\mathbb{S}_{\alpha,1}(t) u_{01}(x), \ldots, \mathbb{S}_{\alpha,m}(t) u_{0m}(x)\right),$$

for all $x \in \mathcal{X}$, $t \in [0,\infty)$. The linear system is decoupled, the operator $\mathbb{S}_\alpha(t)$ acts componentwise; namely, for all $t \in [0,\infty)$, we set

$$\begin{cases} \mathbb{S}_{\alpha,i}(t) = S_{\alpha_i}(t) \equiv I, & \text{for } i = 1, \ldots, r \text{ (when } d_i = 0), \\ \mathbb{S}_{\alpha,i}(t) = S_{\alpha_i}(d_i t)_{|L^\infty(\mathcal{X})}, & \text{for } i = r+1, \ldots, m \text{ (when } d_i > 0). \end{cases}$$

Of course, $r = 0$ is still allowed. In view of Remark 3.1.2, $\mathbb{S}_\alpha(t)$ is not strongly continuous in the Banach space $L^\infty(\mathcal{X}, \mathbb{R}^m)$, which we equip with the canonical sup-norm

$$\|u\|_\infty = \max_{1 \le i \le m} \|u_i\|_{L^\infty(\mathcal{X})}.$$

However, owing to the statement of Proposition 2.2.1, we have $\|\mathbb{S}_\alpha(t) u_0\|_\infty \le \|u_0\|_\infty$, for all $u_0 \in L^\infty(\mathcal{X}, \mathbb{R}^m)$ and $t \in [0,\infty)$. In order to deal with the full nonlinear system, we also define the operator $(\mathbb{P}_\alpha(t) f)$, for $f = (f_1, \ldots, f_m) \in \mathbb{R}^m$, also acting componentwise:

$$\mathbb{P}_\alpha(t) f = \left(\mathbb{P}_{\alpha,1}(t) f_1, \ldots, \mathbb{P}_{\alpha,m}(t) f_m\right),$$

where, for $i = r+1, \ldots, m$,

$$\mathbb{P}_{\alpha,i}(t) f_i \equiv P_{\alpha_i}(d_i t) f_i = \alpha_i t^{\alpha_i - 1} \int_0^\infty \tau \Phi_{\alpha_i}(\tau) S_i(\tau (d_i t)^{\alpha_i}) f_i d\tau$$

and $\mathbb{P}_{\alpha,i}(t) f_i \equiv \left(g_{\alpha_i} * f_i\right)(t)$, for $i = 1, \ldots, r$, in the case of non-diffusing components. Of course, we keep the convention that when some $\alpha_i \equiv 1$ for $i = r+1, \ldots, m$, we let $S_{\alpha_i} \equiv S_i$ and $P_{\alpha_i} \equiv S_i$.

Our notion of mild solution for (4.1.1)–(4.1.2) is stronger than the notion of mild solution from Definition 3.1.3. For the sake of notational convenience, we again let $u(t) = u(\cdot, t)$ and $f(t, u(t)) = f(\cdot, t, u(\cdot, t))$.

Definition 4.1.1 Let $T > 0$ be given, but otherwise arbitrary (and, possibly, $T = \infty$) and let $u_0 \in L^\infty(\mathcal{X}, \mathbb{R}^m)$. We say $u \in E_{\infty,0,T}$ is a mild solution of problem (4.1.1)–(4.1.2) on the time interval $[0,T)$ if:

(a) $u : (x,t) \in \mathcal{X} \times (0,T) \mapsto u(x,t) \in \mathbb{R}^m$ is measurable and $u(\cdot,t) \in L^\infty(\mathcal{X},\mathbb{R}^m)$ such that

$$\sup_{s\in(0,t)} \|u(\cdot,s)\|_\infty =: U < \infty, \text{ for all } t \in (0,T).$$

(b) $u(t) = \mathbb{S}_\alpha(t)u_0 + \int_0^t \mathbb{P}_\alpha(t-\tau) f(\tau, u(\tau))\,d\tau$, for all $t \in (0,T)$, where the integral is an absolutely converging Bochner integral in $L^\infty(\mathcal{X},\mathbb{R}^m)$.

(c) u satisfies the initial condition in the following sense:

$$\lim_{t\to 0^+} \|u(t) - \mathbb{S}_\alpha(t)u_0\|_\infty = 0.$$

We have the first result of this section.

Theorem 4.1.2 (Existence of Maximal Bounded Mild Solutions) *Let assumption* ***(HA)*** *be verified for each operator A_i $(i = 1,\ldots,m)$ and the conditions of* ***(SF1)*** *for the nonlinearity f. For any given $u_0 \in L^\infty(\mathcal{X},\mathbb{R}^m)$, there exists a time $T \in (0,\infty]$ such that the initial-value problem* (4.1.1)–(4.1.2) *possesses a unique mild solution in the sense of Definition 4.1.1 on the interval* $[0,T)$. *Furthermore, the existence time $T \in (0,\infty]$ can be chosen maximal (i.e., the previous statement does not hold for a larger time). In that case, $T = T_{\max}$ and*

$$\lim_{t\to T_{\max}} \|u(t)\|_\infty = \infty, \text{ if } T_{\max} < \infty.$$

Proof Consider the following

$$\begin{aligned}&\textbf{Case1}: \text{all } \alpha_i \equiv 1, i = 1,\ldots,m.\\ &\textbf{Case2}: \text{at least one } \alpha_i \in (0,1), i = 1,\ldots,m.\end{aligned}$$

Let $U_0 \in [0,\infty)$ be such that $\|u_0\|_\infty \le U_0$. Choose $U > U_0$, $T_0 \in (0,1]$ arbitrarily and define the bounded set $B := \overline{\mathcal{X}} \times [0,T_0] \times [-U,U]^m$. Let $L(B) > 0$ be the constant from assumption (**SF1**) and choose $T \in (0,T_0]$ such that

$$U_0 + \widetilde{\theta}(T) \le U, \tag{4.1.3}$$

where the function $\widetilde{\theta}$ is defined as follows:

$$\begin{cases} \widetilde{\theta}(t) := e^{L(B)t} - 1, & \text{in case 1,} \\ \widetilde{\theta}(t) := E_{\alpha_{m_0},1}\left(\dfrac{\Gamma(\alpha_{m_0})L(B)}{\Gamma(\alpha_{M_0})} t^{\alpha_{m_0}}\right) - 1, & \text{in case 2.} \end{cases} \tag{4.1.4}$$

Here, we have set $\alpha_{m_0} := \min_{1\le i\le m}(\alpha_i) \in (0,1)$ and $\alpha_{M_0} := \max_{1\le i\le m}(\alpha_i) \in (0,1]$, and we recall that $E_{\kappa,\beta}(z)$ $(\kappa > 0, \beta \in \mathbb{C})$ is the generalized Mittag-Leffler function

$$E_{\kappa,\beta}(z)=\sum_{n=0}^{\infty}\frac{z^n}{\Gamma(\kappa n+\beta)},\ k\in(0,1).$$

For initial data $u_0\in\mathcal{L}^{\infty}(\mathcal{X},\mathbb{R}^m)$ with $\|u_0\|_{\infty}\leq U_0$, we define the sequence $u^{(l)}$ in $L^{\infty}(\mathcal{X}\times[0,T],\mathbb{R}^m)$, by

$$\begin{cases} u^{(1)}(t)=\mathbb{S}_{\alpha}(t)\,u_0, \\ u^{(l+1)}(t)=\mathbb{S}_{\alpha}(t)\,u_0+\int_0^t\mathbb{P}_{\alpha}(t-s)\,f\left(s,u^{(l)}(s)\right)ds, \end{cases}$$

for $l\in\mathbb{N}$ and $t\in[0,T]$. We claim by induction that the following (4.1.5)–(4.1.9) are satisfied for all $t\in[0,T]$, and all $l\in\mathbb{N}$. We have the estimates:

$$\begin{aligned}&\left\|\left(u^{(l+1)}-u^{(l)}\right)(t)\right\|_{\infty} \\ &\leq L(B)\max_{1\leq i\leq m}\frac{1}{\Gamma(\alpha_i)}\int_0^t(t-\tau)^{\alpha_i-1}\left\|\left(u^{(l)}-u^{(l-1)}\right)(\tau)\right\|_{\infty}d\tau\end{aligned} \tag{4.1.5}$$

and

$$\left\|\left(u^{(l+1)}-u^{(l)}\right)(t)\right\|_{\infty}=\max_{1\leq i\leq m}q_l^i(t)\leq\theta_l(t), \tag{4.1.6}$$

where

$$\begin{cases}\theta_l(t):=\frac{(L(B)t)^l}{l!}, & \text{in case 1,} \\ \theta_l(t):=\frac{1}{\Gamma(l\alpha_{m_0}+1)}\left(\frac{\Gamma(\alpha_{m_0})L(B)}{\Gamma(\alpha_{M_0})}t^{\alpha_{m_0}}\right)^l, & \text{in case 2.}\end{cases} \tag{4.1.7}$$

Also, we have

$$\left\|u^{(l)}(t)\right\|_{\infty}\leq U \tag{4.1.8}$$

and

$$\sum_{1\leq j\leq l}\left\|\left(u^{(j+1)}-u^{(j)}\right)(t)\right\|_{\infty}\leq\widetilde{\theta}(t). \tag{4.1.9}$$

We can easily check these claims for $l = 1$ on account of the definition for $u^{(l)}$, the properties of $\mathbb{S}_\alpha(t)$, $\mathbb{P}_\alpha(t)$ and the conditions of (**SF1**). For instance,

$$\left\| u^{(1)}(t) \right\|_\infty = \|\mathbb{S}_\alpha(t) u_0\|_\infty \le \|u_0\|_\infty < U;$$

since $\|P_{\alpha_i}(t)\|_{L^\infty(\mathcal{X})} \le t^{\alpha_i - 1}/\Gamma(\alpha_i)$, for $t \in (0, T]$, we also get

$$\begin{aligned} q_2^1(t) &:= \left\| u_i^{(2)}(t) - u_i^{(1)}(t) \right\|_{L^\infty} \\ &\le \int_0^t \|P_{\alpha_i}(t-\tau)\|_{\infty,\infty} \left\| f_i\left(\tau, u^{(1)}(\tau)\right) \right\|_{L^\infty(\mathcal{X})} d\tau \\ &\le \frac{L(B)}{\Gamma(\alpha_i)} \int_0^t (t-\tau)^{\alpha_i - 1} d\tau \le \theta_1(t). \end{aligned}$$

Indeed, the Gamma function $\Gamma(x) > 0$ is non-increasing on $(0, \mu]$ and non-decreasing on $[\mu, \infty)$, for some $\mu \in (1, 2)$ (in fact, $\mu = 1.46163..$). This yields that $\Gamma\left(\alpha_{M_0}\right) \le \Gamma(\alpha_i) \le \Gamma\left(\alpha_{m_0}\right)$ in the second case, as well as $\max_{1 \le i \le m} t^{\alpha_i} \le t^{\alpha_{m_0}}$, for $t \in [0, T] \subseteq [0, 1]$; henceforth, we deduce

$$\left\| u^{(2)}(t) - u^{(1)}(t) \right\|_\infty = \max_{1 \le i \le m} q_2^i(t) \le \theta_1(t).$$

Suppose now (4.1.5)–(4.1.9) are already known for some $l - 1 \in \mathbb{N}$. We have to prove these assertions for $l \in \mathbb{N}$. Inequality (4.1.5) is immediate owing to the second condition of (**SF1**), while the assertions (4.1.6) and (4.1.9) are also immediate in the first case by merely performing explicit integration and recalling the series for the exponential function. We now show the same assertions in the second case when at least one of $\alpha_i \in (0, 1)$. By the second condition of (**SF1**), and a change of variable $rt = s$ we have

$$q_l^i(t) \le \int_0^t \|P_{\alpha_i}(t-s)\|_{\infty,\infty} \left\| f\left(s, u^{(l)}(s)\right) - f\left(s, u^{(l-1)}(s)\right) \right\|_{L^\infty(\mathcal{X})} ds \tag{4.1.10}$$

$$\begin{aligned} &\le \frac{L(B)}{\Gamma\left(\alpha_{M_0}\right)} \int_0^t (t-s)^{\alpha_i - 1} \theta_{l-1}(s)\, ds \\ &= \left(\frac{L(B)}{\Gamma\left(\alpha_{M_0}\right)} \right)^l \frac{\Gamma\left(\alpha_{m_0}\right)^{l-1} t^{\alpha_i + (l-1)\alpha_{m_0}}}{\Gamma\left((l-1)\alpha_{m_0} + 1\right)} \mathbb{B}\left(\alpha_i, (l-1)\alpha_{m_0} + 1\right), \end{aligned}$$

where

$$\mathbb{B}(x, y) = \int_0^1 (1-r)^{x-1} r^{y-1} dr$$

is the standard (symmetric) Beta function. Since

$$\mathbb{B}\Big(\alpha_i, (l-1)\,\alpha_{m_0}+1\Big) \leq \mathbb{B}\Big(\alpha_{m_0}, (l-1)\,\alpha_{m_0}+1\Big)$$

and

$$\mathbb{B}\Big(\alpha_{m_0}, (l-1)\,\alpha_{m_0}+1\Big) = \Gamma\left(\alpha_{m_0}\right)\frac{\Gamma\left((l-1)\,\alpha_{m_0}+1\right)}{\Gamma\left(l\alpha_{m_0}+1\right)}$$

we deduce from (4.1.10) that

$$q_l^i\,(t) \leq \theta_l\,(t)\,, \text{ for all } i = 1, \ldots, m,$$

for all $t \in [0, T] \subseteq [0, 1]$. This proves (4.1.6) for any $l \in \mathbb{N}$; (4.1.9) follows from the series of the Mittag-Leffler function $E_{\alpha_{m_0},1}$. It remains to show (4.1.8). Using the definition of the sequence $u^{(l)}$, we get

$$\begin{aligned}\left\|u^{(l)}\,(t)\right\|_\infty &\leq \left\|u^{(1)}\,(t)\right\|_\infty + \sum_{1\leq j<l}\left\|\left(u^{(j+1)}-u^{(j)}\right)(t)\right\|_\infty \\ &\leq U_0 + \widetilde{\theta}\,(t) \leq U,\end{aligned}$$

owing to (4.1.3). Therefore, (4.1.5)–(4.1.9) holds for all $l \in \mathbb{N}$. It follows that there exists a function $u \in L^\infty\left(\mathcal{X}\times[0,T]\,;\mathbb{R}^m\right)$ such that

$$\sup_{t\in[0,T]}\left\|\left(u^{(l)}-u\right)(t)\right\|_\infty \leq \sum_{l\leq j<\infty}\left\|\left(u^{(j+1)}-u^{(j)}\right)(T)\right\|_\infty \to 0,$$

as $l \to \infty$. It is now straightforward to show that the limit u is a solution of our initial-value problem (4.1.1)–(4.1.2) on the time interval $[0, T]$. Since this interval is determined uniformly for all $u_0 \in L^\infty\left(\mathcal{X};\mathbb{R}^m\right)$ such that $\|u_0\|_\infty \leq U_0$, we also have

$$\inf\left\{T\,(u_0) : u_0 \in L^\infty\left(\mathcal{X};\mathbb{R}^m\right),\ \|u_0\|_\infty \leq U_0\right\} > 0, \tag{4.1.11}$$

for all $U_0 \in [0, \infty)$. We prove the second part in the statement of the theorem. We argue by contradiction. Suppose now that there exists $U_0 \in (0, \infty)$ and a sequence $t_n > 0$ such that

$$\lim_{n\to\infty} t_n = T_{\max} < \infty \text{ and } \sup_{n\in\mathbb{N}}\|u\,(t_n)\|_\infty \leq U_0.$$

Hence by (4.1.11), there exists a number $\tau \in (0, \infty)$ and mild solutions $v_n : (x, t) \in \mathcal{X}\times[t_n, t_n+\tau) \mapsto v_n\,(x, t) \in \mathbb{R}^m$ of problem (4.1.1) for an initial datum $u\,(t_n)$ on the interval $[t_n, t_n + \tau)$. Hence by uniqueness, we get a mild solution u for the initial datum u_0 on the larger interval $[0, T_{\max} + \tau)$, which is a contradiction. This completes the proof. □

If the initial datum is sufficiently regular, the mild solution becomes a strong one on any time interval $[T_0, T_{\max})$, for any $T_0 > 0$.

Theorem 4.1.3 (Existence of Maximal Strong Solutions) *Let assumption* **(HA)** *for each operator A_i $(i = 1, \dots, m)$ and the conditions of* **(SF1)–(SF2)** *for the nonlinearity f be satisfied. Assume $u_{0i} \in L^{\infty}(\mathcal{X})$ for $i = 1, \dots, r$ (of course, $r = 0$ is allowed) and $u_{0i} \in D\left(A_{i,p_i}\right) \subset L^{\infty}(\mathcal{X})$ with $p_i \in \left(\beta_{A_i}, \infty\right) \cap (1, \infty)$ for $i = r+1, \dots, m$. Then every bounded mild solution of Theorem 4.1.2 satisfies*

$$u \in C^{0,\kappa}\left([0, T_{\max}); L^{\infty}\left(\mathcal{X}, \mathbb{R}^m\right)\right) \tag{4.1.12}$$

and

$$g_{1-\alpha_i} * (u_i - u_i(0)) \in C^{1,\gamma}\left([0, T_{\max}); L^{\infty}(\mathcal{X})\right),\ i = 1, \dots, r, \tag{4.1.13}$$

$$g_{1-\alpha_i} * (u_i - u_i(0)) \in C^{1}\left((0, T_{\max}); L^{\infty}(\mathcal{X})\right),\ i = r+1, \dots, m, \tag{4.1.14}$$

$$u_i \in C((0, T_{\max}); D(A_{i,p_i})),\ i = r+1, \dots, m, \tag{4.1.15}$$

for some $\kappa, \gamma > 0$. The bounded mild solution also satisfies the initial-value problem (4.1.1)–(4.1.2) *in a strong sense, namely, all equations are satisfied for $t \in (0, T_{\max})$ and almost all $x \in \mathcal{X}$, and*

$$\lim_{t \to 0^+} \|u(t) - u_0\|_{L^{\infty}(\mathcal{X})} = 0. \tag{4.1.16}$$

Proof We define for $i = 1, \dots, m$, $v_i(x,t) = \mathbb{S}_{\alpha,i}(t) u_{0i}$, $G_i(x,t) = f_i(x,t,u(x,t)) \in \mathbb{R}$, for $(x,t) \in \mathcal{X} \times (0, T_{\max})$. The integral solution for the mild solution can be written, with the usual convention, for $i = 1, \dots, m$,

$$u_i(t) = v_i(t) + \int_0^t P_{\alpha_i}(t-s) G_i(s)\, ds, \text{ for } t \in (0, T_{\max}).$$

We argue separately for the diffusing and nondiffusing components. Let $i = r + 1, \dots, m$, and recall that each A_i generates an analytic semigroup $S_i \left(= S_{i,p_i}\right)$ in the space $L^{p_i}(\mathcal{X})$. Let $T \in [T_0, T_{\max})$ be arbitrary for any $T_0 > 0$ and let $0 < T_0 \le t < t + h \le T$ in the estimates below. We first show in what sense the initial datum is satisfied. We first have

$$\begin{aligned} \|u(t) - \mathbb{S}_{\alpha}(t) u_0\|_{\infty} &= \max_{1 \le i \le m} \left\|u_i(t) - \mathbb{S}_{\alpha,i}(t) u_{0i}\right\|_{L^{\infty}(\mathcal{X})} \\ &\le \max_{1 \le i \le m} \int_0^t \left\|P_{\alpha_i}(t-s)\right\|_{\infty,\infty} \|f_i(s, u(s))\|_{\infty}\, ds \end{aligned} \tag{4.1.17}$$

$$\leq L(U) \max_{1\leq i\leq m} \frac{1}{\Gamma(\alpha_i)} \int_0^t (t-s)^{\alpha_i - 1} ds$$

$$\leq L(U) \max_{1\leq i\leq m} \frac{t^{\alpha_i}}{\Gamma(\alpha_i)\alpha_i} \to 0,$$

as $t \to 0^+$. Next, choose $\theta_i > \eta_i$, $\theta_i, \eta_i \in (\beta_{A_i}/p_i, 1)$ such that $\theta_i = \eta_i + \mu_i$. Recall that we have (by Step 2 of the proof of Theorem 3.2.2),

$$t^{1-\alpha_i} \left\| (-A_{i,p_i})^{\eta_i} P_{\alpha_i}(t) \right\|_{p,\infty} \leq C_i t^{-\eta_i \alpha_i}$$

as well as

$$\left\| (-A_{i,p_i})^{-(1-\eta_i)} \left(S_{\alpha_i}(t) - I \right) \right\|_{p_i,p_i} \leq C_i t^{\alpha_i(1-\eta_i)}.$$

Thus, we can argue as in Step 2 of the proof of Theorem 3.2.2 (see, in particular (3.2.16)–(3.2.18) by letting $\overline{q} = 1$, $\chi + 1/\overline{q} = \alpha_i(1-\theta_i)$) to deduce

$$\begin{aligned} \|u_i(t) - u_{0i}\|_{L^\infty(\mathcal{X})} &\leq C \left\| (-A_{i,p_i})^{\theta_i} (u_i(t) - u_{0i}) \right\|_{L^{p_i}(\mathcal{X})} \\ &\leq C t^{\alpha_i(1-\theta_i)} \left(\|u_{0i}\|_{D(A_{i,p_i})} + L(U) \right), \end{aligned} \tag{4.1.18}$$

for some $C > 0$, independent of t, which clearly shows (4.1.16) for the diffusing components. We also need to prove that u_i is continuous with respect to the time variable. As in the proof of Theorem 3.2.2 ($p := p_i$, $A_p := A_{i,p_i}$, $q_1 = q_2 = \infty$, and so forth), see (3.2.19)–(3.2.31), we get

$$\|u_i(t+h) - u_i(t)\|_{L^\infty(\mathcal{X})} \leq C_T L(U) h^\kappa, \ i = r+1, \ldots, m,$$

for some $\kappa > 0$ which depends on $\alpha_i, \beta_{A_i}, \eta_i$. This yields (4.1.12) for the diffusing components, namely

$$u_i \in C^{0,\kappa}\left([0, T_{\max}); L^\infty(\mathcal{X})\right), \ i = r+1, \ldots, m. \tag{4.1.19}$$

Now we consider the non-diffusing components. Let $w = (u_1, \ldots, u_r)$ be the vector of non-diffusing components and define

$$H : (x,t,w) \in X \times [0, T_{\max}) \times \mathbb{R}^r \mapsto H(x,t,w) = f(x,t,w,u_{r+1}, \ldots, u_m) \in \mathbb{R}^r$$

with $f = (f_1, \ldots, f_r)$. The components $i = 1, \ldots, r$, of the integral solution for the mild solution then yields

$$w_i(x,t) = w_{0i}(x) + \int_0^t g_{\alpha_i}(t-s) H_i(x,s,w(x,s))\, ds. \tag{4.1.20}$$

The assumption (**SF2**) together with (4.1.19) implies that

$$|H(x,t,\xi) - H(x,s,\eta)| \leq L_T(U)\left(|t-s|^{\gamma} + |\xi - \eta|\right), \tag{4.1.21}$$

for all $(x,t,\xi), (x,s,\eta) \in \overline{\mathcal{X}} \times [0,T] \times [-U,U]^r$ (the existence of $U > 0$ follows by construction of the mild solution u). Then we deduce for $t \geq \tau \geq 0$ and $i = 1,\ldots,r$,

$$\begin{aligned}&\|w_i(x,t) - w_i(x,\tau)\|_{L^\infty(\mathcal{X})}\\ &\leq \frac{L_T(U)}{\Gamma(\alpha_i)}\int_\tau^t (t-s)^{\alpha_i - 1}\,ds = \frac{L_T(U)(t-\tau)^{\alpha_i}}{\Gamma(\alpha_i+1)}\end{aligned}$$

and so $w_i \in C^{0,\alpha_i}([0,T_{\max}); L^\infty(\mathcal{X}))$, which together with (4.1.19) yields (4.1.12). The foregoing inequality also implies that

$$\lim_{t\to 0^+}\|w_i(t) - w_{0i}\|_{L^\infty(\mathcal{X})} = 0, \; i = 1,\ldots,r.$$

Recalling once again (4.1.19) and the Hölder-Lipschitz condition (4.1.21), we easily infer that $H_i \in C^{0,\gamma}([0,T_{\max}); L^\infty(\mathcal{X}))$, $i = 1,\ldots,r$, for some $\gamma > 0$. Hence, for $i = 1,\ldots,r$, by (4.1.20) and the fact that $g_{1-\alpha} * g_\alpha = 1$, it follows that

$$\partial_t^{\alpha_i} w_i(t) = H_i(\cdot,t,w) \in C^{0,\gamma}([0,T_{\max}); L^\infty(\mathcal{X})). \tag{4.1.22}$$

We finally get the first of (4.1.13). To prove (4.1.14)–(4.1.15) and the remaining part of the statement of the theorem, we argue in a similar fashion as in Step 3 of the proof of Theorem 3.2.2. Thus, the theorem is proved. □

4.2 The Fractional Volterra–Lotka Model

We assume Ω is a bounded domain with Lipschitz continuous boundary $\partial\Omega$ and consider a predator-prey model which assumes a fractional version of the mass action law for the interaction of the two species, predator and prey. As usual, denote the density of prey by $u = u(x,t)$ and of the predator by $v = v(x,t)$. The dynamics of the prey-predator interaction is governed by the following system of reaction–diffusion equations

$$\begin{cases}\partial_t^\alpha u + D_u(-\Delta)_\Omega^s u = u(f - bv), & (x,t) \in \Omega \times (0,\infty),\\ \partial_t v + D_v(-\Delta)_\Omega^l v = v(-g + au), & (x,t) \in \Omega \times (0,\infty),\end{cases} \tag{4.2.1}$$

subject to the following set of boundary conditions

$$\mathcal{N}^{2-2s}u = \mathcal{N}^{2-2l}v = 0 \text{ on } \partial\Omega \times (0, \infty), \tag{4.2.2}$$

and initial conditions

$$(u, v)_{|t=0} = (u_0, v_0) \text{ in } \Omega. \tag{4.2.3}$$

We refer to Appendix B below for a complete description of the boundary operator appearing in (4.2.2). In general a, b, f, g are assumed to be positive constants. Following a discussion in [10], a more general situation can be considered such as, an inhomogeneous environment, symbiosis and saturation can be included by letting the sources f, g depends on x and u, v. We shall consider this situation in a forthcoming article. We assume diffusion rates $D_u, D_v \in (0, \infty)$ and consider the case $s, l \in (1/2, 1)$ since the boundary conditions (4.2.2) makes sense only in this case. Indeed, following an accumulation of evidence from a variety of experimental, theoretical, and field studies [6, 9] we observe that both diffusion operators $(-\Delta)^s_\Omega$, $(-\Delta)^l_\Omega$ offer a better foraging mechanism, than the classical counterpart of Laplacian Δ, for the movement of animals around their natural habitat (cf. also Appendix C, part I). When $s, l \in (0, 1/2]$ and/or $s, l \in \{1\}$, the subsequent results also hold with some minor modifications and different boundary conditions than in (4.2.2) (see Sect. 2.3, for many other possible examples of diffusion operators). The boundary conditions (4.2.2) play a similar role as in the case of no-flux Neumann boundary conditions in that both populations of predator and prey cannot penetrate the boundary $\partial\Omega$. Indeed, we recall that each unforced equation of (4.2.1)–(4.2.2) corresponds to a reflected Lévy process forced to stay inside Ω (see, for instance, [1–3, 5, 7, 8, 12] for the probabilistic point of view). The possible occurrence of a nonlocal derivative ∂_t^α, $\alpha \in (0, 1]$ in the first equation of (4.2.1) accounts for possible effects due to processes with time delay (i.e., "trapping" due environmental and/or predatory effects) in the population of prey (see, for instance, Appendix C.1).

Let $\mathbb{A}_{s,2}$ and $\mathbb{A}_{l,2}$ be the operators on $L^2(\Omega)$ associated with the closed forms

$$\mathcal{E}_s(\varphi, \phi) = \frac{C_{N,s}}{2} D_u \int_\Omega \int_\Omega \frac{(\varphi(x) - \varphi(y))(\phi(x) - \phi(y))}{|x - y|^{N+2s}} dxdy, \quad \varphi, \phi \in W^{s,2}(\Omega),$$

and

$$\mathcal{E}_l(\varphi, \phi) = \frac{C_{N,l}}{2} D_v \int_\Omega \int_\Omega \frac{(\varphi(x) - \varphi(y))(\phi(x) - \phi(y))}{|x - y|^{N+2l}} dxdy, \quad \varphi, \phi \in W^{l,2}(\Omega),$$

respectively. Using the Green formula (B.0.7) we can show that $\mathbb{A}_{s,2}$ and $\mathbb{A}_{l,2}$ are realizations in $L^2(\Omega)$ of $D_u(-\Delta)^s_\Omega$ and $D_v(-\Delta)^l_\Omega$ with the fractional Neumann boundary conditions $\mathcal{N}^{2-2s}\varphi = 0$ on $\partial\Omega$ and $\mathcal{N}^{2-2l}\varphi = 0$ on $\partial\Omega$, respectively.

More precisely we have that

$$\begin{cases} D(\mathbb{A}_{s,2}) = \Big\{\varphi \in W^{s,2}(\Omega):\ D_u(-\Delta)_\Omega^s \varphi \in L^2(\Omega),\ \mathcal{N}^{2-2s}u = 0 \text{ on } \partial\Omega\}, \\ \mathbb{A}_{s,2}\varphi = D_u(-\Delta)_\Omega^s \varphi. \end{cases}$$

and

$$\begin{cases} D(\mathbb{A}_{l,2}) = \{\varphi \in W^{l,2}(\Omega),\ D_v(-\Delta)_\Omega^l \varphi \in L^2(\Omega),\ \mathcal{N}^{2-2l}u = 0 \text{ on } \partial\Omega\}, \\ \mathbb{A}_{l,2}\varphi = D_v(-\Delta)_\Omega^l \varphi. \end{cases}$$

It follows from [4] (cf. also Sect. 2.3) that the operators $A_s := A_{s,2} = -\mathbb{A}_{s,2}$ and $A_l := A_{l,2} := \mathbb{A}_{l,2}$ satisfy the assumption **(HA)**. Throughout the following for $p \in [1,\infty]$ we shall denote by $A_{s,p}$ and $A_{l,p}$ the generator of the associated semigroup on $L^p(\Omega)$. For $p \geq 2$, each such generator possesses in fact the explicit characterization (2.2.10). In addition we shall let

$$\mathcal{L}_s^\infty(\Omega) := \overline{D(A_{s,\infty})}^{L^\infty(\Omega)} \quad \text{and} \quad \mathcal{L}_l^\infty(\Omega) := \overline{D(A_{l,\infty})}^{L^\infty(\Omega)}.$$

We have the following existence result of global strong solutions in the sense introduced in the previous section (see Theorem 4.1.3).

Theorem 4.2.1 *Let* $1/2 < s, l < 1$, $\beta_{A_s} := N/(2s)$ *and* $\beta_{A_l} := N/(2l)$. *Take initial data* $u_0 \in D(A_{s,p_s}) \subset L^\infty(\Omega)$, $v_0 \in D(A_{l,p_l}) \subset L^\infty(\Omega)$ *for some* $p_s \in (\beta_{A_s},\infty) \cap (1,\infty)$, $p_l \in (\beta_{A_l},\infty) \cap (1,\infty)$ *such that* $u_0 \geq 0$, $v_0 \geq 0$. *Then the system* (4.2.1)–(4.2.3) *has a unique global strong solution* $u \geq 0$, $v \geq 0$ *on the time interval* $(0,\infty)$ *satisfying*

$$\lim_{t\to 0} \|u(t) - u_0\|_{L^\infty(\Omega)} = 0, \tag{4.2.4}$$

$$\lim_{t\to 0} \|v(t) - v_0\|_{L^\infty(\Omega)} = 0. \tag{4.2.5}$$

In addition for every $T \in (0,\infty)$ *the following estimates hold:*

$$\sup_{0<t<T} \|u(t)\|_{L^\infty(\Omega)} < \infty, \tag{4.2.6}$$

$$\sup_{0<t<T} \|v(t)\|_{L^\infty(\Omega)} < \infty. \tag{4.2.7}$$

Proof Let u_0 and v_0 be as in the statement of the theorem. Recall that the operator $A_{s,2}$ and $A_{l,2}$ satisfy the assumption **(HA)**. Define the function $F : \Omega \times [0,\infty) \times \mathbb{R}^2 \to \mathbb{R}^2$ by $F(x,t,(\xi,\eta)) := \Big(\xi(f - b\eta), \eta(-g + a\xi)\Big)$. Then F satisfies the assumptions **(SF1)** and **(SF2)** with $\gamma = 1$. It follows from Theorem 4.1.3 that the system (4.2.1)–(4.2.3) has a strong solution (u,v) on some maximal interval

$[0, T_{\max})$. The strong solution is given by the integral representation

$$u(t) = S_{u,\alpha}(t)u_0 + \int_0^t P_{u,\alpha}(t-\tau)\,(u(f-bv))\,(\tau)d\tau, \tag{4.2.8}$$

$$v(t) = S_v(t)v_0 + \int_0^t S_v(t-\tau)\,(v(-g+au))\,(\tau)d\tau, \tag{4.2.9}$$

where $S_{u,\alpha}(t)$, $S_v(t)$ denote the resolvent family and semigroup on $L^2(\Omega)$ generated by the operators $A_{s,2}$ and $A_{l,2}$, respectively. Here we have also defined

$$S_{u,\alpha}\,(t)\,\omega := \int_0^\infty \Phi_\alpha(\tau)S_u(\tau t^\alpha)\omega d\tau, \;\; P_{u,\alpha}(t)\omega := \alpha t^{\alpha-1}\int_0^\infty \tau\Phi_\alpha(\tau)S_u(\tau t^\alpha)\omega d\tau$$

and set $P_{u,1}(t) \equiv S_{u,1}\,(t) := S_u\,(t)$. Recall that $u_0 \geq 0$ and $v_0 \geq 0$ on Ω and define the functions $k, h : \Omega \times (0,\infty) \times \mathbb{R} \to \mathbb{R}$ by $k(x,t,\xi) = \xi(f - bv(x,t))$ and $h(x,t,\xi) = \xi(-g + au(x,t))$. Since $k(x,t,0) = h(x,t,0) = 0$, it follows from Theorem 3.6.1 that $u(x,t) \geq 0$ and $v(x,t) \geq 0$ for a.e. $(x,t) \in \Omega \times [0, T_{\max})$.

Note that each component of F satisfies the assumption (**F6**) for $u \geq 0$, $v \geq 0$. Applying Theorem 3.4.11, we get that there exist two constants $C, C_0 > 0$ such that for every $0 < T < \infty$,

$$\sup_{0<t<T} \|u(t)\|_{L^\infty(\Omega)} \leq C\left(\|u_0\|_{L^\infty(\Omega)} + E_{\alpha,1}\,(C_0T) + T^\alpha E_{\alpha,1}\,(C_0T)\,f\right) \tag{4.2.10}$$

and we have shown (4.2.6) for any $\alpha \in (0,1]$. Next we consider (4.2.9) and let $p(x,t,\xi) = (-g + au)$. Since g is a positive constant and $u \geq 0$, we have

$$h\,(x,t,\xi)\,\xi = p(x,t,\xi)\xi^2 = (-g+au)\xi^2 \leq au(x,t)\xi^2.$$

It follows from (4.2.10) that $c_0 := \sup_{t\in(0,T)} \|au(x,t)\|_{L^\infty(\Omega)} < \infty$. We have shown that h also satisfies the assumption (**F6**). Then applying Theorem 3.4.11 once again and recalling Corollary 3.4.14, we get that there there exist two constants $C, C_0 > 0$ such that for every $0 < T < \infty$,

$$\sup_{t\in(0,T)} \|v(t)\|_{L^\infty(\Omega)} \leq C\left(\|v_0\|_{L^\infty(\Omega)} + c_0\,(T+1)\,e^{C_0T}\right). \tag{4.2.11}$$

We have shown (4.2.7). Together with (4.2.6) we can conclude that $T_{\max} = \infty$ (see Sect. 4.1). For (4.2.4)–(4.2.5), we refer once again to the proof of Theorem 4.1.3 (see (4.1.18), (4.1.17) and set $\alpha_1 = \alpha$, $\alpha_2 = 1$, in which case $P_{\alpha_1} \equiv P_{u,\alpha}$, $P_{\alpha_2} \equiv S_v$); they easily follow now on the account of (4.2.10)–(4.2.11). The proof is finished. □

We recall from (3.1.4) that $\mathfrak{n}_\mathfrak{s} := \beta_{A_s}\alpha$, $\alpha \in (0,1]$ and $\mathfrak{n}_\mathfrak{l} := \beta_{A_l}$.

Theorem 4.2.2 *Let $p_0, q_0 \in [1, \infty]$ such that $\beta_{A_s}/p_0 < 1$ and consider initial data $0 \leq u_0 \in L^{p_0}(\Omega)$, $0 \leq v_0 \in L^{q_0}(\Omega)$. Then the fractional Lotka-Volterra system* (4.2.1)–(4.2.3) *has a unique global mild solution $u \geq 0$, $v \geq 0$ on the time interval $[0, \infty)$, given by* (4.2.8)–(4.2.9)*, which is also a strong solution on $(0, \infty)$. Moreover, the pair (u, v) satisfies*

$$\lim_{t \to 0^+} \|u(t) - u_0\|_{L^1(\Omega)} = 0, \quad \lim_{t \to 0^+} \|v(t) - v_0\|_{L^k(\Omega)} = 0, \text{ for } k \in [1, q_0), \tag{4.2.12}$$

and the following estimates, for any $T \in (0, \infty)$:

$$\sup_{t \in (0,T)} (t \wedge 1)^{\delta_s} \|u(t)\|_{L^p(\Omega)} < \infty, \, p \in [p_0, \infty], \tag{4.2.13}$$

$$\sup_{t \in (0,T)} (t \wedge 1)^{\delta_l} \|v(t)\|_{L^q(\Omega)} < \infty, \, q \in [q_0, \infty], \tag{4.2.14}$$

where

$$\delta_s := \frac{\mathfrak{n}_s}{p_0}\left(1 - \frac{p_0}{p}\right),$$

$$\delta_l := \frac{\mathfrak{n}_l}{q_0}\left(1 - \frac{q_0}{q}\right).$$

The proof of the theorem follows from a series of propositions and lemmas that we subsequently give. In what follows one can start with more regular initial data due to the statement of Theorem 4.2.1 and then deduce all the required estimates with less regular initial data by exploiting a standard approximation argument.

Proposition 4.2.3 *Every nonnegative solution (u, v) satisfies the following estimate*[1]

$$\|u(t)\|_{L^{p_0}(\Omega)} \leq \|u_0\|_{L^{p_0}(\Omega)} \left(E_{\alpha,1}\left(f t^\alpha\right)\right)^{\frac{1}{p_0}}, \tag{4.2.15}$$

for all $t \in (0, \infty)$ and $\alpha \in (0, 1]$, $p_0 \in [1, \infty)$.

Proof We derive the estimate in case $p_0 \in (1, \infty)$, the cases $p_0 \in \{1, \infty\}$ follow directly from a limit argument in (4.2.15). Multiply the first equation of (4.2.1) by $p_0 u^{p_0 - 1}$, integrate the resulting identity over Ω, then exploit the first inequality of Proposition 3.4.9 if $\alpha \in (0, 1)$ and use the fact that $u, v \geq 0$. We find that

$$\partial_t^\alpha \left(\|u(t)\|_{L^{p_0}(\Omega)}^{p_0}\right) \leq f \|u(t)\|_{L^{p_0}(\Omega)}^{p_0},$$

[1] The Mittag-Leffler function $E_{1,1}(x) = e^x$.

for all $t \geq 0$. The comparison principle of Lemma A.0.7 then immediately yields the result since the unique solution of $\partial_t^\alpha y = fy$ is $y = y(0) E_{\alpha,1}(ft^\alpha)$. □

Lemma 4.2.4 *Let $p_0 \in [1, \infty]$ such that $\beta_{A_s}/p_0 < 1$ and assume sufficiently smooth data (u_0, v_0). Then for every $T \in (0, \infty)$ there exists a constant $M = M(T, \Omega, a, b, f, g) \in (0, \infty)$ such that*

$$\sup_{t\in(0,T)} (t \wedge 1)^{\delta_s} \|u(t)\|_{L^p(\Omega)} \leq M, \ p \in [p_0, \infty]. \tag{4.2.16}$$

Proof *We apply Lemma 3.4.3 to the equation in u and use the one-sided version due to Remark 3.4.5 since $u, v \geq 0$. The weight function $c(x,t) = f$ is constant, we have $q_1 = q_2 = r_2 = \infty$, $r_1 = p_0$ and $\gamma = 1$ and we can find a number $\tilde{b} \in [0, 1)$ satisfying $\left(1-\tilde{b}\right)\frac{\beta_{A_s}\alpha}{p_0} < \alpha - \varepsilon$, for some $\varepsilon \in (0, \alpha)$. Note that $\beta_{A_s}/p_0 < 1$ is equivalent to $\beta_{A_s}\alpha/p_0 = \mathfrak{n}_\mathfrak{s}/p_0 < \alpha$. The assertion (3.4.18) of Lemma 3.4.3 then implies the existence of a constant $C_* > 0$ independent of u_0, u, U, t and T such that*

$$\|u(\cdot,t)\|_{L^\infty(\Omega)} \leq C_*(t \wedge 1)^{-\frac{\mathfrak{n}_\mathfrak{s}}{p_0}}\left[\|u_0\|_{L^{p_0}(\Omega)} + \Upsilon(t)\left(U + U^{1/\left(1-\tilde{b}\right)}\right)\right], \tag{4.2.17}$$

for all $t \in (0, T]$. This yields (4.2.16) for $p = \infty$ since by the definition of U, c and (4.2.15), we have

$$U := f\,\|1 + |u|\|_{p_0,\infty,T}^{\left(1-\tilde{b}\right)} < \infty.$$

In case $p = p_0$, estimate (4.2.16) is just the a priori estimate (4.2.15), namely, it follows that

$$\sup_{t\in(0,T)} \|u(t)\|_{L^{p_0}(\Omega)} \leq M_1(T, f) < \infty. \tag{4.2.18}$$

Since both $L^{p_0}(\Omega)$ and $L^\infty(\Omega)$-estimates are now readily available by (4.2.18) and (4.2.17), we can use the interpolation inequality

$$\|u\|_{L^p(\Omega)} \leq \|u\|_{L^{p_0}(\Omega)}^{p_0/p} \|u\|_{L^\infty(\Omega)}^{1-p_0/p}, \ p \in [p_0, \infty], \tag{4.2.19}$$

to derive the desired estimate in (4.2.16) for arbitrary p. □

Lemma 4.2.5 *Under the assumptions of Lemma 4.2.4, every nonnegative solution v of (4.2.1)–(4.2.3) satisfies*

$$\sup_{t\in(0,T)} \|v(t)\|_{L^{q_0}(\Omega)} < \infty, \ \textit{for } q_0 \in [1, \infty]. \tag{4.2.20}$$

Proof Consider the weight function $c(x,t) = -g + au(x,t)$ and notice that $v \geq 0$ is a solution of

$$\partial_t v + \mathbb{A}_{l,2} v = cv,\ v(0) = v_0.$$

Multiply this equation by qv^{q-1} and integrate the resulting identity over Ω. Since

$$\left(\mathbb{A}_{l,2} v, v^{q-1}\right)_{L^2(\Omega)} \geq 0$$

by (2.2.11), we find

$$\partial_t \|v(t)\|^q_{L^q(\Omega)} \leq q \|v(t)\|^q_{L^q(\Omega)} \|c(t,\cdot)\|_{L^\infty(\Omega)},$$

for all $t \in [0, T]$. This inequality implies that

$$\partial_t \|v(t)\|_{L^q(\Omega)} \leq \|v(t)\|_{L^q(\Omega)} \|c(t,\cdot)\|_{L^\infty(\Omega)}$$

and the application of Gronwall's inequality yields

$$\|v(t)\|_{L^q(\Omega)} \leq \|v_0\|_{L^q(\Omega)} e^{\int_0^t \|c(s,\cdot)\|_{L^\infty(\Omega)} ds}, \tag{4.2.21}$$

for any $q \in [1, \infty)$. Notice that in view of (4.2.17), $\|c(t,\cdot)\|_{L^\infty(\Omega)} \sim t^{-\mathfrak{n}_{\mathfrak{s}}/p_0}$ for $t \in (0,1)$ and $\|c(t,\cdot)\|_{L^\infty(\Omega)} \leq C_T$ for $t \geq 1$; thus we have $\|c(t,\cdot)\|_{L^\infty(\Omega)} \in L^1(0,T)$ since $\mathfrak{n}_{\mathfrak{s}}/p_0 < \alpha \leq 1$ by assumption. In particular, we infer from (4.2.21) and (4.2.17) the existence of a constant $M = M(T, p_0, \mathfrak{n}_{\mathfrak{s}}, g, a, f, u_0) > 0$, independent of q, such that

$$\|v(t)\|_{L^q(\Omega)} \leq M \|v_0\|_{L^q(\Omega)},\ t \in [0, T],$$

which is exactly the primary estimate (4.2.20) for $q = q_0 \in [1, \infty)$. Passing to the limit as $q \to \infty$ in the previous inequality, we also get the estimate (4.2.20) for $q_0 = \infty$. We thus conclude the proof. □

We can now show that v satisfies the smoothing property (4.2.14).

Lemma 4.2.6 *Under the assumptions of Lemma 4.2.4, for every $T \in (0, \infty)$ there exists a constant $M = M(T, \Omega, a, b, f, g) \in (0, \infty)$ such that*

$$\sup_{t \in (0,T)} (t \wedge 1)^{\delta_l} \|v(t)\|_{L^q(\Omega)} \leq M,\ q \in [q_0, \infty]. \tag{4.2.22}$$

Proof We first notice that estimate (4.2.16) with $p = \infty$ implies that $\|u\|_{\infty, q_2} \leq M$, for some $q_2 \in (1, 1/\delta)$, $\delta := \mathfrak{n}_{\mathfrak{s}}/p_0 < \alpha \leq 1$. This time we apply Lemma 3.4.3 to the equation in v with the weight function $c(x,t) = -g + au(x,t)$ which now satisfies $\|c\|_{\infty, q_2} \leq M_1$, and set $q_1 = r_2 = \infty$, $q_2 := q_2 \in (1, 1/\delta)$, $r_1 = q_0$

and $\gamma = 1$. The constant $M_1 \in (0, \infty)$ depends on the final time $T > 0$ but is independent of t. Indeed, we can find a *new* number $\tilde{b} \in [0, 1)$, sufficiently close to 1, satisfying

$$\frac{1}{q_2} + \left(1 - \tilde{b}\right) \frac{\mathfrak{n}_{\mathfrak{l}}}{q_0} < 1 - \varepsilon,$$

for some $\varepsilon \in (0, 1)$. It follows from the assertion (3.4.18) of Lemma 3.4.3 that there exists a constant $C_* > 0$ independent of v_0, v, V and t such that

$$\|v(\cdot, t)\|_{L^\infty(\Omega)} \leq C_* (t \wedge 1)^{-\frac{\mathfrak{n}_{\mathfrak{l}}}{q_0}} \left[\|u_0\|_{L^{q_0}(\Omega)} + \Upsilon(t) \left(V + V^{1/\left(1-\tilde{b}\right)} \right) \right], \tag{4.2.23}$$

for all $t \in (0, T]$. Here, $V < \infty$ is defined as

$$V := \|1 + |v|\|_{q_0,\infty,T}^{\left(1-\tilde{b}\right)} \|c\|_{\infty,q_2}.$$

This yields estimate (4.2.22) for $q = \infty$. Next, recall that v also satisfies (4.2.20); this allows us to exploit an interpolation similar to (4.2.19) in the spaces $L^\infty(\Omega) \subset L^q(\Omega) \subset L^{q_0}(\Omega)$. Thus we arrive at the desired estimate (4.2.14) for an arbitrary $q \in [q_0, \infty]$ and we conclude the proof. □

Proposition 4.2.7 *Assume $p_0, q_0 \in [1, \infty]$ are such that $\beta_{A_s}/p_0 < 1$. Then the following assertions hold.*

(a) *There exists a constant $M > 0$ such that for every $t \in (0, 1)$, we have*

$$\left\|u(t) - S_{u,\alpha}(t) u_0\right\|_{L^{q_0}(\Omega)} \leq M t^{\varepsilon}, \quad \|v(t) - S_v(t) v_0\|_{L^{q_0}(\Omega)} \leq M t^{\varepsilon}, \tag{4.2.24}$$

for some $\varepsilon > 0$, for $p_0, q_0 \in [1, \infty)$.

(b) *For $i = 1, 2$, let (u_i, v_i) be a solution of* (4.2.1)–(4.2.3) *corresponding to an initial datum (u_{0i}, v_{0i}). Then for every $T \in (0, \infty)$, there exists a constant $C = C(T) \in (0, \infty)$, independent of (u_i, v_i), such that*

$$\begin{aligned} &|||u_1 - u_2|||_{\infty,\delta,T} + |||v_1 - v_2|||_{q_0,0,T} \\ &\leq C \left(\|u_{01} - u_{02}\|_{L^{p_0}(\Omega)} + \|v_{01} - v_{02}\|_{L^{q_0}(\Omega)} \right). \end{aligned} \tag{4.2.25}$$

Proof By the integral formula (4.2.8) and estimate (4.2.13) with $p = \infty$ and $\delta = \mathfrak{n}_{\mathfrak{s}}/p_0$, for $t \in (0, 1)$ we have as in (3.1.16) (with $s_0 := q_0$, $p_0 := q_0$,

$q_2 := \infty$, $q_1 := q_0$),

$$\left\| u(t) - S_{u,\alpha}(t) u_0 \right\|_{L^{q_0}(\Omega)} \leq C \int_0^t (\tau \wedge 1)^{-\delta} d\tau \left(|||u|||_{\infty,\delta,1} \right) \left(1 + |||v|||_{q_0,0,1} \right)$$
$$\leq M t^{\varepsilon},$$

for some $\varepsilon > 0$. The same argument applied to the difference $v(t) - S_v(t) v_0$ in (4.2.9) gives the required estimate in (4.2.24).

In order to show (4.2.25), we take $\varepsilon := 1 - \delta > 0$, where $\delta = \mathfrak{n}_{\mathfrak{s}} / p_0$. Subtracting the integral equations (4.2.8) corresponding to each $i = 1, 2$ and u_i, we obtain

$$\| u_1(t) - u_2(t) \|_{L^\infty(\Omega)} \leq \left\| S_{u,\alpha}(t) \right\|_{\infty, p_0} \| u_{01} - u_{02} \|_{L^{p_0}(\Omega)} \tag{4.2.26}$$
$$+ \Lambda(u_i, v_i) \int_0^t C \left\| P_{u,\alpha}(t-s) \right\|_{\infty, q_0} (s \wedge 1)^{-\delta} ds,$$

where

$$\Lambda(u_i, v_i) := |||u_1 - u_2|||_{\infty,\delta,T} \left(1 + |||v_1|||_{q_0,0,T} \right) \tag{4.2.27}$$
$$+ \left(1 + |||u_2|||_{\infty,\delta,T} \right) |||v_1 - v_2|||_{q_0,0,T} \,.$$

We can apply Lemma A.0.1 to the second summand in (4.2.26) and exploit the global bounds (4.2.13)–(4.2.14) to estimate the corresponding norms for u_2 and v_1. We deduce

$$|||u_1 - u_2|||_{\infty,\delta,t} \tag{4.2.28}$$
$$\leq C \| u_{01} - u_{02} \|_{L^{p_0}(\Omega)}$$
$$+ C_T \Upsilon(t) \left(|||u_1 - u_2|||_{\infty,\delta,T} + |||v_1 - v_2|||_{q_0,0,T} \right),$$

for all $t \in (0, T]$, for some $C > 0$ independent of t. Arguing similarly for the v-component, we find

$$\| v_1(t) - v_2(t) \|_{L^{q_0}(\Omega)} \leq \| S_v(t) \|_{q_0,q_0} \| v_{01} - v_{02} \|_{L^{q_0}(\Omega)}$$
$$+ C_T \Lambda(u_i, v_i) \int_0^t (s \wedge 1)^{-\delta} ds$$

which yields

$$|||v_1 - v_2|||_{q_0,0,t} \leq C \| v_{01} - v_{02} \|_{L^{q_0}(\Omega)} \tag{4.2.29}$$
$$+ C_T \Upsilon(t) \left(|||u_1 - u_2|||_{\infty,\delta,T} + |||v_1 - v_2|||_{q_0,0,T} \right),$$

for all $t \in (0, T]$. Choose now a small enough $h > 0$ such that $C_T \Upsilon(h) \leq 1/2$ into (4.2.28)–(4.2.29). We obtain

$$|||u_1 - u_2|||_{\infty,\delta,h} + |||v_1 - v_2|||_{q_0,0,h} \leq M(T)\left(\|u_{01} - u_{02}\|_{L^{p_0}(\Omega)} + \|v_{01} - v_{02}\|_{L^{q_0}(\Omega)}\right). \tag{4.2.30}$$

With the same proof, we can also infer that

$$\begin{aligned} &|||(u_1 - u_2)(\cdot, t_0 + \cdot)|||_{\infty,\delta,h} + |||(v_1 - v_2)(\cdot, t_0 + \cdot)|||_{q_0,0,h} \\ &\leq M(T)\left(\|(u_1 - u_2)(t_0)\|_{L^{p_0}(\Omega)} + \|(v_1 - v_2)(t_0)\|_{L^{q_0}(\Omega)}\right), \end{aligned} \tag{4.2.31}$$

for all $t_0 \in [0, T)$. We can now apply the estimate (4.2.31) successively for $j = 0, 1, 2, \ldots$, with initial data $(u, v)(t_0 + jh)$. Then the assertion (4.2.25) follows by induction on j and we finish the proof of the proposition. □

Proof (Proof of Theorem 4.2.2) The proof follows now by a simple procedure where we approximate any rough nonnegative initial data $(u_0, v_0) \in L^{p_0}(\Omega) \times L^{q_0}(\Omega)$ by a sequence of nonnegative functions $(u_{0n}, v_{0n}) \in D(A_{s,p_s}) \times D(A_{l,p_l})$ (for some sufficiently large $p_s \in (\beta_{A_s}, \infty)$, $p_l \in (\beta_{A_l}, \infty)$ and $p_s, p_l \geq 2$) such that

$$\|u_{0n} - u_0\|_{L^{p_0}(\Omega)} \to 0, \; \|v_{0n} - v_0\|_{L^{q_0}(\Omega)} \to 0, \text{ as } n \to \infty$$

with

$$\|u_{0n}\|_{L^{p_0}(\Omega)} \leq \|u_0\|_{L^{p_0}(\Omega)}, \quad \|v_{0n}\|_{L^{q_0}(\Omega)} \leq \|v_0\|_{L^{q_0}(\Omega)}.$$

The above lemmata and propositions then hold with the constants $M, M_1, C, C_* > 0$ independent of n for the sequence of strong solutions (u_n, v_n). Thus, assertion (4.2.25) of Proposition 4.2.7 implies that the sequence (u_n, v_n) converges to (u, v) in $E_{\infty,\delta,T} \times E_{q_0,0,T}$, and all the a priori estimates derived in this section also hold for the limit solution (u, v). It is then straightforward to show from (4.2.8)–(4.2.9) that (u, v) is also the mild solution of system (4.2.1)–(4.2.3) for an initial datum $(u_0, v_0) \in L^{p_0}(\Omega) \times L^{q_0}(\Omega)$ (see Chap. 3 and Sect. 4.1). In particular every such mild solution (u, v) is global and bounded on $[T_0, \infty)$ for every $T_0 > 0$, and one can use arguments as in the proofs of Theorems 4.1.3 and 3.2.6, respectively, to show that (u, v) is also a strong solution on $[2T_0, \infty)$. The continuity properties in (4.2.12) follow also immediately by virtue of (4.2.24) and Remark 3.1.2. □

4.3 A Fractional Nuclear Reactor Model

Let $\Omega \subset \mathbb{R}^N$ be a bounded domain with Lipschitz continuous boundary $\partial\Omega$ and consider the following parabolic system as a prototype for a nuclear reactor model that we believe has a more realistic physical interpretation than the classical one (see Appendix C). Let $u = u(x,t)$ represent the fast neutron density and $v = v(x,t)$ be the fuel temperature at any point $x \in \Omega$ and for any time $t \geq 0$. The system for (u, v) reads

$$\begin{cases} \partial_t^\alpha u + (-\Delta)_\Omega^s u = u(\lambda - bv), & (x,t) \in \Omega \times (0,\infty), \\ \partial_t^\beta v = -cv + au, & (x,t) \in \Omega \times (0,\infty), \end{cases} \tag{4.3.1}$$

subject to the following set of boundary and initial conditions:

$$\mathcal{N}^{2-2s} u = 0 \text{ on } \partial\Omega \times (0,\infty), \quad (u,v)_{|t=0} = (u_0, v_0) \text{ in } \Omega. \tag{4.3.2}$$

Here $s \in (1/2, 1)$, $\alpha, \beta \in (0,1)$ and λ, a, b, c are positive constants in the model equations (4.3.1)–(4.3.2) and $(-\Delta)_\Omega^s$ is the regional fractional Laplace operator in Ω (see (2.3.19)) and $\mathcal{N}^{2-2s}$ denotes the corresponding fractional Neumann derivative (see Sect. 2.3). The first (unforced) equation of (4.3.1) may be derived from a continuous-time random walk with temporal memory (see Appendix C.3), while incorporating avalanche-like transport effects in the neutron density and the second equation can be analogously derived on similar principles as those considered in Appendix C, by ignoring any diffusion effects in the fluid temperature v. We note that the case $s = \alpha = \beta = 1$ has been treated by Rothe [10] in some detail as a simple reactor model proposed in [11].

Note that the first equation of (4.3.1) is structurally the same as the equation for prey in the fractional Lotka-Volterra model investigated in Sect. 4.2. Thus the arguments appear to be even more simple than in that case provided that we can derive suitable a priori estimates for the fluid temperature in (4.3.1). Let $S_u(t)$ denote the semigroup on $L^2(\Omega)$ generated by the operator $A_{s,2}$, as given previously. Consider a sufficiently smooth initial datum (u_0, v_0) and its corresponding solution.

Proposition 4.3.1 *The fluid temperature v satisfies the following estimate*

$$\sup_{t\in(0,T)} \|v(t)\|_{L^q(\Omega)} \leq C^{1/q} \max\left\{ \|v_0\|_{L^q(\Omega)}, \frac{\varepsilon^{1/q-1}}{(C_\varepsilon q)^{1/\alpha q}} \sup_{t\in(0,T)} \|u(t)\|_{L^\infty(\Omega)} \right\}, \tag{4.3.3}$$

for some $\varepsilon > 0$ depending only on a, c, and some constants $C, C_\varepsilon > 0$ independent of $q \in [1,\infty]$.

Proof By application of Proposition 3.4.9 into the second equation of (4.3.1), we get

$$\begin{aligned}
&\partial_t^\beta \left(\|v(t)\|_{L^q(\Omega)}^q \right) + cq \|v(t)\|_{L^q(\Omega)}^q \\
&\leq aq \|v(t)\|_{L^q(\Omega)}^{q-1} \|u(t)\|_{L^\infty(\Omega)} \\
&\leq a\varepsilon^{1-q} \left(\sup_{t\in(0,T)} \|u(t)\|_{L^\infty(\Omega)} \right)^q + a\varepsilon(q-1) \|v(t)\|_{L^q(\Omega)}^q ,
\end{aligned}$$

for all $t \geq 0$. This inequality implies for a sufficiently small $\varepsilon \in (0, ca/2]$ and $C_\varepsilon = c/2$, that

$$\begin{aligned}
&\partial_t^\beta \left(\|v(t)\|_{L^q(\Omega)}^q \right) + C_\varepsilon q \|v(t)\|_{L^q(\Omega)}^q \\
&\leq M := a \left(\varepsilon^{1/q-1} \sup_{t\in(0,T)} \|u(t)\|_{L^\infty(\Omega)} \right)^q .
\end{aligned}$$

We infer by Lemma A.0.8 the existence of a constant $C > 0$, independent of q, such that

$$\|v(t)\|_{L^q(\Omega)}^q \leq C \max \left\{ \|v_0\|_{L^q(\Omega)}^q , \frac{a}{(C_\varepsilon q)^{1/q}} \left(\varepsilon^{1/q-1} \sup_{t\in(0,T)} \|u(t)\|_{L^\infty(\Omega)} \right)^q \right\} .$$

Taking the $1/q$-root on both sides, this inequality gives the desired assertion in (4.3.3) for every $q \in [1, \infty)$. Since the constants C, C_ε involved in (4.3.3) are independent of q, we also recover the estimate in case $q = \infty$, by passing to the limit as $q \to \infty$ in (4.3.3). □

In view of the simple estimate of Proposition 4.3.1, we can derive the existence of unique global strong solution in the sense of Theorem 4.1.3.

Theorem 4.3.2 *Let $1/2 < s < 1$ and $\beta_{A_s} := N/(2s)$. Take initial data $u_0 \in D(A_{s,p_s}) \subset L^\infty(\Omega)$, $v_0 \in L^\infty(\Omega)$ for some $p_s \in (\beta_{A_s}, \infty) \cap (1, \infty)$ such that $u_0 \geq 0$, $v_0 \geq 0$. Then the system (4.3.1)–(4.3.2) has a unique global strong solution $u \geq 0$, $v \geq 0$ on the time interval $(0, \infty)$ satisfying*

$$\lim_{t\to 0} \|u(t) - u_0\|_{L^\infty(\Omega)} = 0, \quad \lim_{t\to 0} \|v(t) - v_0\|_{L^\infty(\Omega)} = 0. \tag{4.3.4}$$

In addition for every $T \in (0, \infty)$ the following estimates hold:

$$\sup_{0<t<T} \|u(t)\|_{L^\infty(\Omega)} < \infty, \quad \sup_{0<t<T} \|v(t)\|_{L^\infty(\Omega)} < \infty. \tag{4.3.5}$$

Proof Let u_0 and v_0 be as in the statement of the theorem. We can infer the existence of a maximally defined strong solution by Theorem 4.1.3, given as

$$u(t) = S_{u,\alpha}(t)u_0 + \int_0^t P_{u,\alpha}(t-\tau)\,(u(\lambda - bv))\,(\tau)d\tau, \tag{4.3.6}$$

$$v(t) = v_0 + \int_0^t g_\beta\,(t-\tau)\,(-cv + au)\,(\tau)d\tau, \tag{4.3.7}$$

for $t \in (0, T_{\max})$. A similar argument to the proof of Theorem 4.2.1 successively yields that $u(x,t) \geq 0$ and then $v(x,t) \geq 0$ for a.e. $(x,t) \in \Omega \times [0, T_{\max})$ since $g\,(u, 0) = au \geq 0$ (for $g\,(u, v) := -cv + au$). Moreover, the first bound of (4.3.5) is satisfied by the same arguments of Theorem 4.2.1. Consequently, so is the second bound of (4.3.5) on account of (4.3.3) in case $q = \infty$. The continuity properties in (4.3.4) follow also by similar arguments on account of (4.3.5) with the exception that for the integral solution v we have a more direct estimate from (4.3.7). The proof is finished. □

As a consequence of Proposition 4.2.3 and Lemma 4.2.4 we immediately have the following estimate since the equation for u is the same as for the fractional system (4.2.1)–(4.2.2).

Proposition 4.3.3 *Let $p_0 \in [1, \infty]$ such that $\beta_{A_s}/p_0 < 1$ and assume a sufficiently smooth datum u_0. Then for every $T \in (0, \infty)$ there exists a constant $M = M\,(T, \Omega, b, \lambda) \in (0, \infty)$ such that*

$$\sup_{t \in (0,T)} (t \wedge 1)^{\delta_s} \, \|u\,(t)\|_{L^p(\Omega)} \leq M, \;\; p \in [p_0, \infty]\,, \tag{4.3.8}$$

where $\delta_s \geq 0$ is as in the statement of Theorem 4.2.2.

We now derive some uniform a priori L^q-estimate for the temperature. Of course, there is no smoothing effect in the component v other than the one implied by u. In other words, v turns out to be as regular as u but no more. Since $u, v \geq 0$, we have by (4.3.7) that pointwise in time,

$$v\,(t) \leq \overline{v}\,(t) := v_0 + a \int_0^t g_\beta\,(t-\tau)\,u(\tau)d\tau \tag{4.3.9}$$

and so it suffices to derive the required estimate for $\overline{v}$.

Proposition 4.3.4 *Under the assumptions of Proposition 4.3.3, it holds for any $v_0 \in L^q\,(\Omega)$, $q \leq p$ with $p \in [p_0, \infty]$ and $T \in (0, \infty)$, the estimate*

$$\sup_{t \in (0,T)} \|v\,(t)\|_{L^q(\Omega)} < \infty \text{ if } \beta \geq \delta_s$$

and

$$\sup_{t\in(0,T)} (t\wedge 1)^{\delta_s-\beta} \|v(t)\|_{L^q(\Omega)} < \infty \text{ if } \beta < \delta_s.$$

Proof We have

$$\|\overline{v}(t)\|_{L^q(\Omega)} \le \|v_0\|_{L^q(\Omega)} + a\int_0^t g_\beta(t-\tau)\|u(\tau)\|_{L^q(\Omega)}\,d\tau \quad (4.3.10)$$

$$\le \|v_0\|_{L^q(\Omega)} + C\left(\sup_{t\in(0,T)} (t\wedge 1)^{\delta_s}\|u(t)\|_{L^p(\Omega)}\right)$$

$$\times \int_0^t g_\beta(t-\tau)(\tau\wedge 1)^{-\delta_s}\,d\tau.$$

A basic change of variable $s=\tau/t$ gives for $t<1$,

$$\int_0^t g_\beta(t-\tau)(\tau\wedge 1)^{-\delta_s}\,d\tau = C_\beta t^{\beta-\delta_s}\int_0^1 s^{-\delta_s}(1-s)^{\beta-1}\,d\tau,$$

where the latter integral is convergent since $\beta>0$ and $\delta_s=\frac{n_s}{p_0}\left(1-\frac{p_0}{p}\right)<\alpha\left(1-\frac{p_0}{p}\right)<\alpha<1$. When $t>1$ we argue as in the proof of Lemma A.0.1 to split the integral over intervals $k<t\le k+1$, such that

$$\int_0^t = \int_0^1 + \int_1^2 + \ldots + \int_{k-2}^{k-1} + \int_{k-1}^{t-1} + \int_{t-1}^t.$$

It follows that

$$\int_0^t g_\beta(t-\tau)(\tau\wedge 1)^{-\delta_s}\,d\tau \le C(k+1)\le 2Ct,$$

for some constant $C>0$ independent of t,T. We then infer the existence of a positive constant $M_2=M_2(M,\Omega,a,T,v_0)\in(0,\infty)$ such that

$$\sup_{t\in(0,T)} (t\wedge 1)^{\delta_s-\beta}\|\overline{v}(t)\|_{L^q(\Omega)} \le M_2, \text{ if } \beta\le\delta_s$$

and

$$\sup_{t\in(0,T)} \|\overline{v}(t)\|_{L^q(\Omega)} \le M_2, \text{ if } \beta>\delta_s.$$

We may now conclude using (4.3.9). □

Proposition 4.3.5 *Let the assumptions of Proposition 4.3.3 be satisfied and let $q_0 \in [1, p_0]$ such that $v_0 \in L^{q_0}(\Omega)$. Then the following estimate holds:*

$$\sup_{t\in(0,T)} \|v(t)\|_{L^{q_0}(\Omega)} \leq \|v_0\|_{L^{q_0}(\Omega)} + C_T, \tag{4.3.11}$$

for some constant $C_T \in (0, \infty)$ that depends only on the $L^{p_0}(\Omega)$-norm of u_0, T and the other physical parameters of the problem.

Proof Let $T \in (0, \infty)$ be arbitrary. The Hölder inequality on the bounded interval $[0, T]$ yields in (4.3.11), owing to the fact that $g_\beta \in L^1(0, T)$,

$$\sup_{t\in(0,T)} \|\overline{v}(t)\|_{L^{q_0}(\Omega)} \leq \|v_0\|_{L^{q_0}(\Omega)} + C_T \|u\|_{L^\infty(0,T;L^{p_0}(\Omega))},$$

for some $C_T = C(a, \Omega, T, \beta) > 0$ independent of t. Application of (4.3.8) with $p = p_0$ then gives the desired estimate in (4.3.11) since $\delta_s = 0$ and $v \leq \overline{v}$. □

Proposition 4.3.6 *Assume $p_0 \in [1, \infty]$ such that $\beta_{A_s}/p_0 < 1$ ($\Leftrightarrow \mathfrak{n}_\mathfrak{s}/p_0 < \alpha$) and $q_0 \in [1, p_0] \cap (\beta_{A_s}, \infty]$. Then the following assertions hold.*

(a) There exists a constant $M > 0$ such that for every $t \in (0, 1)$, we have

$$\left\|u(t) - S_{u,\alpha}(t)u_0\right\|_{L^{p_0}(\Omega)} \leq Mt^\varepsilon, \quad \|v(t) - v_0\|_{L^{q_0}(\Omega)} \leq Mt^\beta, \tag{4.3.12}$$

for some small $\varepsilon > 0$.

(b) For $i = 1, 2$, let (u_i, v_i) be a solution of (4.3.1)–(4.3.2) corresponding to an initial datum (u_{0i}, v_{0i}). Then for every $t \in (0, T)$, there exists a constant $C = C(T) \in (0, \infty)$, independent of (u_i, v_i), such that

$$\begin{aligned} &|||u_1 - u_2|||_{p_0,0,t} + |||v_1 - v_2|||_{q_0,0,t} \\ &\leq C\left(\|u_{01} - u_{02}\|_{L^{p_0}(\Omega)} + \|v_{01} - v_{02}\|_{L^{q_0}(\Omega)}\right). \end{aligned} \tag{4.3.13}$$

Proof We first prove (4.3.12) by following a similar argument that we employed in the proof of Lemma 3.1.5 (see (3.1.16)) by viewing $c(x, t) := \lambda - bv$, $f(x, t, u) = c(x, t)u$, with $q_1 := q_0$, $q_2 := \infty$. To this end, let $T \in (0, 1)$, $0 \leq t \leq T$ and recall the uniform estimates (4.3.8), (4.3.11), which imply that

$$|||c|||_{q_0,0,T} \leq C\,|||1 + v|||_{q_0,0,T} \leq N_1, \quad ||||u||||_{p,\delta_s,T} \leq N_2. \tag{4.3.14}$$

Then let $s_0 \in [1, \infty)$ be such that

$$\delta_s \leq \frac{1}{s_0} \text{ and } \frac{\mathfrak{n}_\mathfrak{s}}{s_0} + \delta_s + \varepsilon < \alpha + \frac{\mathfrak{n}}{p_0},$$

for a sufficiently small $\varepsilon \in (0, \alpha]$ such that $\varepsilon + \delta_s \leq \alpha$. We subsequently apply the statement of Lemma A.0.1 with the choices $p := p_0$, $s_1 := s_0$, $s_2 := \infty$, $\theta := \delta_s$,

$\delta := 0$ and $\varepsilon := \varepsilon$ (note again that $r(\tau) \equiv \|c(\cdot,\tau)\|_{L^{q_0}(\mathcal{X})}$ and $p_{s_2}(r) = \|c\|_{q,\infty}$). Once again if $s_0 \geq p_0$ is arbitrary we have that $\mathfrak{n}_{\mathfrak{s}}/s_0 - \mathfrak{n}_{\mathfrak{s}}/p_0 \in [0, 2\alpha)$ is trivially satisfied, while if $s_0 < p_0$ one may choose s_0 sufficiently close to $p_0 \in [1, \infty)$ such that $1/s_0 < 2/\beta_{A_s} + 1/p_0$. Note that the assumptions of Lemma A.0.1 are satisfied with the above choices of $\delta, s_1, s_2, p, \varepsilon, \theta$, since $0 \leq \delta_s < \alpha < 1$ and $\varepsilon + \delta_s \leq \alpha$, and

$$\frac{\mathfrak{n}_{\mathfrak{s}}}{s_0} < \alpha + \frac{\mathfrak{n}_{\mathfrak{s}}}{p_0}.$$

Indeed, by virtue of Hölder's inequality, for all $t \in (0, T] \subset (0, 1)$ we have

$$\begin{aligned}
&\left\| u(\cdot,t) - S_{u,\alpha} u_0 \right\|_{L^{p_0}(\Omega)} \qquad (4.3.15)\\
&\leq \left(\int_0^t \left\| P_{u,\alpha}(t-\tau) \right\|_{p_0,s_0} (\tau \wedge 1)^{-\delta_s} \|\lambda - bv\|_{L^{q_0}(\mathcal{X})} d\tau \right) |||u|||_{p,\delta_s,T}\\
&\leq C\,|||1 + v|||_{q_0,0,T}\, t^{\varepsilon}\, ||||u||||_{p,\delta_s,T}\\
&\leq C N_1 N_2 t^{\varepsilon}
\end{aligned}$$

owing once again to (4.3.14). This gives the first of the assertion (4.3.12). For the second estimate, by (4.3.7) we have for every $1 \leq q \leq q_0$,

$$\begin{aligned}
\|v(t) - v_0\|_{L^q(\Omega)} &\leq \int_0^t g_\beta(t-\tau) \|(-cv + au)(\tau)\|_{L^q(\Omega)} d\tau \qquad (4.3.16)\\
&\leq C\left(||||v||||_{q_0,0,T} + ||||u||||_{p_0,0,T} \right) \int_0^t g_\beta(t-\tau)\, d\tau\\
&\leq C(N_1 + N_2)\, t^{\beta},
\end{aligned}$$

for all $0 \leq t \leq T < 1$.

Next, we prove the continuous dependence estimate (4.3.13). By virtue of (4.3.14), from (4.2.27) we have the uniform bound

$$\Lambda(u_i, v_i) \leq |||u_1 - u_2|||_{p_0,0,t}(1 + N_1) + (1 + N_2)\,|||v_1 - v_2|||_{q_0,0,t}$$

so that the same argument exploited in (4.3.15) in the integral formulation (4.3.6) for the difference u, yields for $t \in (0, 1)$,

$$||||u_1 - u_2||||_{p_0,0,t} \leq C\,\|u_{01} - u_{02}\|_{L^{p_0}(\Omega)} + C t^{\varepsilon}\, ||||v_1 - v_2||||_{q_0,0,t}\,. \qquad (4.3.17)$$

By (4.3.7), we obtain as in (4.3.16), for $q \leq q_0 \leq p_0$, that

$$\|v_1(t) - v_2(t)\|_{L^q(\Omega)} \qquad (4.3.18)$$

$$\leq \|v_{01} - v_{02}\|_{L^q(\Omega)} + \int_0^t g_\beta (t-\tau) \|(-cv+au)(\tau)\|_{L^q(\Omega)} d\tau$$
$$\leq \|v_{01} - v_{02}\|_{L^q(\Omega)} + Ct^\beta \left(||||v_1 - v_2||||_{q_0,0,t} + ||||u_1 - u_2||||_{p_0,0,t}\right).$$

Define $\rho := \min\{\varepsilon, \beta\} > 0$ and the function

$$\psi(t) := ||||v_1 - v_2||||_{q_0,0,t} + ||||u_1 - u_2||||_{p_0,0,t}.$$

By the estimates (4.3.17)–(4.3.18), for a sufficiently small $t < 1$, it holds

$$\psi(t) \leq C\left(\|u_{01} - u_{02}\|_{L^{p_0}(\Omega)} + \|v_{01} - v_{02}\|_{L^{q_0}(\Omega)}\right) + Ct^\rho \psi(t), \qquad (4.3.19)$$

for some constant $C > 0$ independent of t. Further choose $t_0 \ll 1$ such that $Ct_0^\rho \leq 1/2$ and observe that (4.3.19) also implies

$$\psi(t) \leq 2C\left(\|u_{01} - u_{02}\|_{L^{p_0}(\Omega)} + \|v_{01} - v_{02}\|_{L^{q_0}(\Omega)}\right), \qquad (4.3.20)$$

for all $t \in (0, t_0]$. Finally, we can employ (4.3.20) successively with initial data $(u, v)(t + it_0)$, for $i = 0, 1, 2, \ldots$, since by (4.3.8) and (4.3.11),

$$\sup_{i \in \mathbb{N}} \left(\|u(t + it_0)\|_{L^{p_0}(\Omega)} + \|v(t + it_0)\|_{L^{q_0}(\Omega)}\right) \leq N_3.$$

Indeed, for the same step size t_0, the assertion (4.3.20) yields the estimate

$$\sup_{t \in [t_0 + it_0, t_0 + (i+1)t_0]} \left(||||v_1 - v_2||||_{q_0,0,t} + ||||u_1 - u_2||||_{p_0,0,t}\right) \qquad (4.3.21)$$
$$\leq C\left(\|(u_1 - u_2)(t_0 + it_0)\|_{L^{p_0}(\Omega)} + \|(v_1 - v_2)(t_0 + it_0)\|_{L^{q_0}(\Omega)}\right),$$

for all $i \in \{0, 1, 2, \ldots.\}$. Then the assertion (4.3.13) on the whole interval $(0, T)$ follows by an induction procedure on i, applied successively in (4.3.21). Thus, the proposition is proved. □

We conclude the section with the second result concerning the well-posed problem of mild solutions.

Theorem 4.3.7 *Assume $p_0 \in [1, \infty]$ such that $\beta_{A_s}/p_0 < 1$ and $q_0 \in [1, p_0] \cap (\beta_{A_s}, \infty]$, and let $0 \leq u_0 \in L^{p_0}(\Omega)$, $0 \leq v \in L^{q_0}(\Omega)$ be such that $u_0 \geq 0$ and $v_0 \geq 0$ a.e. on Ω. If $p_0 = \infty$, in addition assume $u_0 \in \mathcal{L}_s^\infty(\Omega)$. Then the fractional system (4.3.1)–(4.3.2) has a unique global mild solution $u \geq 0, v \geq 0$ on the time interval $[0, \infty)$, given by (4.3.6)–(4.3.7), which hold as absolutely convergent Bochner integrals in $L^1(\Omega)$. Moreover, the pair (u, v) satisfies*

$$\lim_{t \to 0^+} \|u(t) - u_0\|_{L^{p_0}(\Omega)} = 0, \quad \lim_{t \to 0^+} \|v(t) - v_0\|_{L^{q_0}(\Omega)} = 0 \qquad (4.3.22)$$

and the uniform estimates stated in Propositions 4.3.3 and 4.3.5.

Proof The proof follows by a standard approximation procedure. The initial datum $(u_0, v_0) \in L^{p_0}(\Omega) \times L^{q_0}(\Omega)$ can be approximated by a convenient sequence of regular initial data (u_{0n}, v_{0n}), according to the statement of Theorem 4.3.2. In particular, this sequence may be chosen such that

$$\lim_{n\to\infty}\left[\|u_{0n} - u_0\|_{L^{p_0}(\Omega)} + \|v_{0n} - v_0\|_{L^{q_0}(\Omega)}\right] = 0 \tag{4.3.23}$$

and

$$\|u_{0n}\|_{L^{p_0}(\Omega)} \le \|u_0\|_{L^{p_0}(\Omega)}, \quad \|v_{0n}\|_{L^{q_0}(\Omega)} \le \|v_0\|_{L^{q_0}(\Omega)}, \tag{4.3.24}$$

for all $n \in \mathbb{N}$. Let now (u_n, v_n) be the global strong solution for an initial datum (u_{0n}, v_{0n}). All the constants occurring in Propositions 4.3.3–4.3.5 can be chosen independent of $n \in \mathbb{N}$, owing to (4.3.24). Furthermore, the assertion (4.3.13) of Proposition 4.3.6, together with (4.3.23), implies that the sequence (u_n, v_n) converges to $(u, v) \in E_{p_0,\delta_s,T} \times E_{q_0,0,T}$, in the sense that

$$||||v_n - v||||_{q_0,0,t} + ||||u_n - u||||_{p_0,0,t} \to 0, \text{ as } n \to \infty,$$

for all $t \in (0, T)$. Besides, all the estimates of Propositions 4.3.3–4.3.5 hold for the limit solution (u, v) as well. By the same arguments as in the proof of Lemma 3.1.5, it is now straightforward to show that (u, v) is indeed the mild solution of the system (4.3.1)–(4.3.2), for any initial datum (u_0, v_0). The conclusion (4.3.22) is also a consequence of Proposition 4.3.6 and Remark 3.1.2. □

Corollary 4.3.8 *The mild solution (u, v) of* (4.3.1)–(4.3.2) *is also regularizing in the sense that its first component u becomes a global strong solution on $[T_0, \infty)$, for every $T_0 > 0$, as well as, the second component $v \in L^\infty([T_0, \infty); L^\infty(\Omega))$.*

References

1. K. Bogdan, K. Burdzy, Z.-Q. Chen, Censored stable processes. Probab. Theory Relat. Fields **127**(1), 89–152 (2003)
2. Z.-Q. Chen, T. Kumagai, Heat kernel estimates for stable-like processes on d-sets. Stoch. Process. Appl. **108**(1), 27–62 (2003)
3. A.A. Dubkov, B. Spagnolo, V.V. Uchaikin, Lévy flight superdiffusion: an introduction. Int. J. Bifurcat. Chaos Appl. Sci. Eng. **18**(9), 2649–2672 (2008)
4. C.G. Gal, M. Warma, Nonlocal transmission problems with fractional diffusion and boundary conditions on non-smooth interfaces. Commun. Partial Differ. Equ. **42**(4), 579–625 (2017)
5. R. Gorenflo, F. Mainardi, A. Vivoli, Continuous-time random walk and parametric subordination in fractional diffusion. Chaos Solitons Fractals **34**(1), 87–103 (2007)
6. N.E. Humphries, N. Queiroz, J.R.M. Dyer, N.G. Pade, M.K. Musyl, K.M. Schaefer, D.W. Fuller, J.M. Brunnschweiler, T.K. Doyle, J.D.R. Houghton et al., Environmental context explains lévy and brownian movement patterns of marine predators. Nature **465**(7301), 1066 (2010)

7. M. Jara, Nonequilibrium scaling limit for a tagged particle in the simple exclusion process with long jumps. Commun. Pure Appl. Math. **62**(2), 198–214 (2009)
8. A. Mellet, S. Mischler, C. Mouhot, Fractional diffusion limit for collisional kinetic equations. Arch. Ration. Mech. Anal. **199**(2), 493–525 (2011)
9. A.M. Reynolds, C.J. Rhodes, The Lévy flight paradigm: random search patterns and mechanisms. Ecology **90**(4), 877–887 (2009)
10. F. Rothe, *Global Solutions of Reaction-Diffusion Systems*. Lecture Notes in Mathematics, vol. 1072 (Springer, Berlin, 1984)
11. E.T. Rumble III, W.E. Kastenberg, On the application of eigenfunction expansions to problems in nonlinear space-time reactor dynamics. Nucl. Sci. Eng. **49**(2), 172–187 (1972)
12. D. Schertzer, M. Larchevêque, J. Duan, V.V. Yanovsky, S. Lovejoy, Fractional Fokker-Planck equation for nonlinear stochastic differential equations driven by non-Gaussian Lévy stable noises. J. Math. Phys. **42**(1), 200–212 (2001)

Chapter 5
Final Remarks and Open Problems

In this monograph, we first consider a semilinear fractional kinetic equation that is characterized by the presence of a nonlinear time-dependent source $f = f(x, t, u)$, a generalized time derivative ∂_t^α in the sense of Caputo and the presence of a large class of diffusion operators A. Many examples of diffusion operators that satisfy our assumptions are given in Sect. 2.3. We give a unified analysis, using tools in semigroups theory and the theory of partial differential equations (Sects. 2.1 and 2.2), in order to obtain sharp results for the well-posedness problem of mild and strong solutions (Sects. 3.1 and 3.2), as well as for the global regularity problem in Sect. 3.4. Further properties, such as nonnegativity of the mild (and/or strong) solutions and their limiting behavior as $\alpha \to 1$, are also provided in Sects. 3.6 and 3.5, respectively. Finally, in Sect. 3.7 an application of these results is given.

The framework we develop for the scalar equation in Chap. 3 is then extended in the second part of the monograph (Chap. 4) to nonlinear systems of fractional kinetic equations. Here, we first develop a general scheme that allows to establish sharp results for the well-posedness problem of (locally-defined) mild and strong solutions associated with such general systems (Sect. 4.1). We then combine this analysis with that of the previous chapters to derive well-posedness results in terms of globally defined mild and strong solutions, for a fractional prey-predator model (Sect. 4.2) and a simple fractional nuclear reaction model (Sect. 4.3). In addition, we provide a number of important technical tools in Appendix A, in support of the analysis developed in this monograph; this appendix is followed by Appendix B, which contains several results concerning the regional fractional Laplace operator associated with fractional Neumann and/or Robin boundary conditions. Finally, in Appendix C, we recall the current scientific literature for different kinds of fractional kinetic equations that are suggested by concrete problems in mathematical physics, probability and finance, and which fully motivated the analysis in this monograph.

We give next a number of final comments and discuss possible open problems.

C. G. Gal, M. Warma, *Fractional-in-Time Semilinear Parabolic Equations and Applications*, Mathématiques et Applications 84,
https://doi.org/10.1007/978-3-030-45043-4_5

Remark 5.0.1 Our main working hypothesis in this monograph is that the underlying physical space $\mathcal{X}$ is a (relatively) compact Hausdorff space. However, we note that this assumption has been placed just for the sake of technical convenience. Much of the results developed in Chap. 3 are true for instance when $\mathcal{X}$ is only locally compact (say when $\mathcal{X}$ is replaced by either $\mathbb{R}^N$, or half-space $\mathbb{R}_+^N$ or an unbounded open set $\Omega \subset \mathbb{R}^N$). Indeed, all the supporting technical results given in Appendix A, with the exception of Lemma A.0.2, are still valid when $\mathcal{X}$ is only locally compact. In particular, it means that the results on well-posedness of (locally-defined) mild and strong solutions are still true in that case, with the exception of case (c) of Theorem 3.1.4; we recall that this case uses Lemma A.0.2 in a crucial way. Moreover, one may obtain the same global bounds derived in Sect. 3.4 by making proper modifications in the proofs when $\mathcal{X}$ is only locally compact.

Problem 1 Prove the analogue of Lemma A.0.2 when $\mathcal{X}$ *is only locally compact.*

Remark 5.0.2 Let us consider the semilinear parabolic problem (3.1.1) with the nonlinearity $f(x,t,u) = c(x,t)|u|^{\gamma-1}u$, for some $c \in L_{q_1,q_2}$. We note that the critical exponent γ, as stated by Theorem 3.1.4,

$$\frac{\mathfrak{n}}{q_1} + \frac{1}{q_2} + (\gamma - 1)\frac{\mathfrak{n}}{p_0} \leq \alpha, \ \mathfrak{n} := \beta_A \alpha, \ \alpha \in (0,1], \tag{5.0.1}$$

is in fact optimal in the sense that there are always locally-defined mild solutions for some $u_0 \in L^{p_0}(\mathcal{X})$. When instead $\gamma \geq 1$ and $p_0 \geq 1$ satisfy the inequality

$$\frac{\mathfrak{n}}{q_1} + \frac{1}{q_2} + (\gamma - 1)\frac{\mathfrak{n}}{p_0} > \alpha, \tag{5.0.2}$$

we conjecture that problem (3.1.1) does not have any locally-defined mild solution for certain initial data $u_0 \in L^{p_0}(\mathcal{X})$. Indeed, this was already discovered by Weissler [12, 13] for the classical problem when $\alpha = 1$, $\beta_A = N/2$ and $q_1 = q_2 = \infty$; (5.0.2) recovers the super-critical range $(\gamma - 1)\frac{N}{2p_0} > 1$ in that case.

Problem 2 Prove the above conjecture in the super-critical case (5.0.2).

Problem 3 Consider the problem (3.1.1) in the subcritical and limiting cases as defined by (5.0.1). Several further open problems can be considered:

(a) Under the same assumptions of Chap. 3, investigate the long-term behavior of (3.1.1) in terms of global attractors and ω-limit sets.

(b) Under proper conditions on the nonlinearity and the diffusion operator A, show that each globally defined solution converges to a unique steady state u_* as time goes to infinity, where u_* is a proper solution of the corresponding stationary problem.

(c) Investigate the blow-up phenomenon for Problem (3.1.1) for various diffusion operators. We refer the reader to [11] when $A = \Delta$.

(d) Give a further refined regularity analysis to show the (Hölder) continuity of solutions for the abstract problem (3.1.1) for a large class of diffusion operators A. We recall that such result has already been proven in [1] for the corresponding problem with $f = f(x, t)$ and A is given by Example 2.3.6(a). When $A = \Delta$ or a second-order operator in divergence form, this has been proven in [14].

Problem 4 The current framework can be extended to accommodate more general transmission problems than the ones considered in [4, 5].

Problem 5 The framework in Sect. 3.3 can be further developed to show higher-order differentiability properties for the strong solution under additional assumptions of the nonlinear function f.

Remark 5.0.3 The framework developed in this monograph can be exploited to obtain global existence of solutions to other interesting reaction–diffusion systems that contain some fractional kinetics. Among them, one can consider more general systems based on ecological interactions and physical models based on chemical reactions with anomalous diffusion that may occur in spatially inhomogeneous media (cf. Appendix C). Among such interesting systems, one may mention the fractional Brusselator for reaction kinetics [9] which was considered in [6] as a physical model for activator-inhibitor dynamics that exhibits anomalous behavior.

Problem 6 Consider the fractional Brusselator discussed by Henry and Wearne [6] and prove the existence of globally-defined strong and mild solutions. This is an open problem in light of the difficulties that arise from the nature of the coupling in the system and the corresponding nonlinear terms. We refer the reader to the survey paper of Pierre [8] for more information regarding the classical reaction–diffusion problem when $\alpha_i \equiv 1, i \in \{1, \ldots, m\}$.

Problem 7 Investigate the long-term behavior of solutions, as time goes to infinity, to the fractional Volterra–Lotka and nuclear reactor systems introduced in Sects. 4.2 and 4.3, respectively.

Remark 5.0.4 The contribution [7] contains an analogue of the classical Aubin-Lion compactness lemma in order to obtain existence of weak solutions to some nonlinear systems that involve a fractional Caputo derivative. This approach can be also applied to the semilinear problem (1.0.1) in order to develop a well-defined L^2-theory. However, our approach doesn't require any compactness arguments and is of more general interest since it is developed in the L^p-setting. Moreover, our theory can be also extended for problems (1.0.1) with notions other than the Caputo fractional derivative for as long as one can provide a formula for the solution similar to (3.1.2). This is in particular very useful in those situations where the integral kernel in the Caputo derivative is slightly more general than $g_{1-\alpha}$ (see (2.1.1)). These issues shall be addressed in future contributions.

To conclude this section we list some open problems regarding the exterior value elliptic problems (Eqs. (2.3.21), (2.3.25) and (2.3.27)) for the fractional Laplace operator. We refer to [3] for more details.

Problem 8 Let $u \in W_0^{s,2}(\overline{\Omega})$ be a weak solution of the Dirichlet exterior value problem (2.3.21). Prove or disprove that u is a strong solution of (2.3.21).

Remark 5.0.5 Assume that $\Omega \subset \mathbb{R}^N$ is a bounded domain of class $C^{1,1}$. Let $(\varphi_n)_{n\geq 0}$ be the eigenfunctions of the operator $(-\Delta)_D^s$ (see Example 2.3.6(a)). It has been shown in [2, Section 5] (see also [10] for the case $N = 1$) that for every $n \geq 1$, $\varphi_n \in C^{0,s}(\overline{\Omega})$ and $\varphi_n \notin C^{0,\gamma}(\overline{\Omega})$ for any $\gamma > s$.

Problem 9 Let $u \in W_\Omega^{s,2}$ be a weak solution of the Neumann exterior value problem (2.3.25). Prove that $u \in C(\mathbb{R}^N)$ and $u|_\Omega \in W_{\text{loc}}^{2s,2}(\Omega)$. Prove or disprove that u is a strong solution of (2.3.25).

Problem 10 Assume that $\Omega \subset \mathbb{R}^N$ is a bounded domain of class $C^{1,1}$. Let $(\psi_n)_{n\geq 0}$ be the eigenfunctions of the operator $(-\Delta)_N^s$ (see Example 2.3.6(b)). Prove that for every $n \geq 1$, $\psi_n \in C^{0,s}(\overline{\Omega})$ and $\psi_n \notin C^{0,\gamma}(\overline{\Omega})$ for any $\gamma > s$.

Problem 11 Let $u \in W_{\beta,\Omega}^{s,2}$ be a weak solution of the Robin exterior value problem (2.3.27). Prove that $u|_\Omega \in W_{\text{loc}}^{2s,2}(\Omega)$. Prove or disprove that u is a strong solution of (2.3.27). Assume that $\beta \in L^1(\mathbb{R}^N \setminus \Omega) \cap L^\infty(\mathbb{R}^N \setminus \Omega)$. Prove that $u \in C(\mathbb{R}^N)$.

Problem 12 Assume that $\Omega \subset \mathbb{R}^N$ is a bounded domain of class $C^{1,1}$ and that $\beta \in C_c^1(\mathbb{R}^N \setminus \Omega)$. Let $(\phi_n)_{n\geq 0}$ be the eigenfunctions of the operator $(-\Delta)_R^s$ (see Example 2.3.6(c)). Prove that for every $n \geq 1$, $\phi_n \in C^{0,s}(\overline{\Omega})$ and $\phi_n \notin C^{0,\gamma}(\overline{\Omega})$ for any $\gamma > s$.

References

1. M. Allen, L. Caffarelli, A. Vasseur, A parabolic problem with a fractional time derivative. Arch. Ration. Mech. Anal. **221**(2), 603–630 (2016)
2. U. Biccari, M. Warma, E. Zuazua, Local elliptic regularity for the Dirichlet fractional Laplacian. Adv. Nonlinear Stud. **17**(2), 387–409 (2017)
3. B. Claus, M. Warma, Realization of the fractional laplacian with nonlocal exterior conditions via forms method. J. Evol. Equ. (2020). https://doi.org/10.1007/s00028-020-00567-0
4. C.G. Gal, M. Warma, Transmission problems with nonlocal boundary conditions and rough dynamic interfaces. Nonlinearity **29**(1), 161–197 (2016)
5. C.G. Gal, M. Warma, Nonlocal transmission problems with fractional diffusion and boundary conditions on non-smooth interfaces. Commun. Partial Differ. Equ. **42**(4), 579–625 (2017)
6. B.I. Henry, S.L. Wearne, Existence of Turing instabilities in a two-species fractional reaction-diffusion system. SIAM J. Appl. Math. **62**(3), 870–887 (2001/2002)
7. L. Li, J.-G. Liu, Some compactness criteria for weak solutions of time fractional PDEs. SIAM J. Math. Anal. **50**, 3693–3995 (2018)

8. M. Pierre, Global existence in reaction-diffusion systems with control of mass: a survey. Milan J. Math. **78**(2), 417–455 (2010)
9. I. Prigogine, R. Lefever, Symmetry breaking instabilities in dissipative systems. II. J. Chem. Phys. **48**(4), 1695–1700 (1968)
10. R. Servadei, E. Valdinoci, On the spectrum of two different fractional operators. Proc. R. Soc. Edinb. Sect. A **144**(4), 831–855 (2014)
11. V. Vergara, R. Zacher, Stability, instability, and blowup for time fractional and other nonlocal in time semilinear subdiffusion equations. J. Evol. Equ. **17**(1), 599–626 (2017)
12. F.B. Weissler, Local existence and nonexistence for semilinear parabolic equations in L^p. Indiana Univ. Math. J. **29**(1), 79–102 (1980)
13. F.B. Weissler, Existence and non-existence of global solutions for a semilinear heat equation. Isr. J. Math. **38**(1–2), 29–40 (1981)
14. R. Zacher, A De Giorgi–Nash type theorem for time fractional diffusion equations. Math. Ann. **356**(1), 99–146 (2013)

Appendix A
Some Supporting Technical Tools

We first state a result that gives an estimate on time convolution integrals involving the ultra-contractive bounded operator $P_\alpha(t)$, $\alpha \in (0, 1]$.

Lemma A.0.1 *Define* $\mathfrak{n} = \beta_A \alpha > 0$. *Let* $p, s_1 \in [1, \infty]$ *such that* $\mathfrak{n}(1/s_1 - 1/p) < 2\alpha$ *and* $s_2 \in (1/\alpha, \infty]$, $\theta \geq 0$, $\varepsilon \in [0, \alpha)$, $\delta \in [0, \infty)$ *satisfy*

$$\frac{\mathfrak{n}}{s_1} + \frac{1}{s_2} < \alpha + \frac{\mathfrak{n}}{p}, \quad \frac{\mathfrak{n}}{s_1} + \frac{1}{s_2} + \theta + \varepsilon \leq \alpha + \frac{\mathfrak{n}}{p} + \delta \tag{A.0.1}$$

and

$$\frac{1}{s_2} + \theta < 1, \quad \frac{1}{s_2} + \theta + \varepsilon \leq \alpha + \delta. \tag{A.0.2}$$

Let $r : [0, \infty) \to r(t) \in \mathbb{R}$ *be a measurable function such that*

$$p_{s_2}(r) := \sup_{t_1, t_2 \in [0,\infty), 0 \leq t_2 - t_1 \leq 1} \left(\int_{t_1}^{t_2} |r(t)|^{s_2} dt \right)^{\frac{1}{s_2}} < \infty.$$

Define the function

$$g(t) := (t \wedge 1)^\delta \int_0^t \|P_\alpha(t - \tau)\|_{p,s_1} (\tau \wedge 1)^{-\theta} r(\tau) d\tau, \ \forall t \geq 0.$$

Then there exists a constant $C > 0$ *independent of* t *such that*

$$|g(t)| \leq C (t \wedge 1)^\varepsilon p_{s_2}(r), \ \textit{for any } t \in (0, 1]$$

C. G. Gal, M. Warma, *Fractional-in-Time Semilinear Parabolic Equations and Applications*, Mathématiques et Applications 84,
https://doi.org/10.1007/978-3-030-45043-4

and

$$|g(t)| \leq tCp_{s_2}(r), \text{for any } t > 1.$$

Proof The idea is to combine the proof of [44, Lemma 6] with the new ultracontractivity estimate of Proposition 2.2.2 (see (b)). To this end, we define $s \in [1, \infty]$ such that $1/s + 1/s_2 = 1$. In what follows, if $p \leq s_1$ we let $\beta = 0$; otherwise if $p > s_1$, we choose $\beta \in (0, 1/s + \alpha - 1)$ such that

$$\frac{\mathfrak{n}}{s_1} - \frac{\mathfrak{n}}{p} < \beta \text{ and } \frac{1}{s_2} + \beta + \theta + \varepsilon \leq \alpha + \delta. \tag{A.0.3}$$

Note that $\beta \in (0, 1/s + \alpha - 1)$ is equivalent to $\beta \in (0, \alpha - 1/s_2)$, where we recall that $s_2 \in (1/\alpha, \infty]$. Also notice that (A.0.3) can be achieved owing to the assumption (A.0.1). By Proposition 2.2.2, part (b), there exists a constant $C > 0$, independent of t, such that

$$\|P_\alpha(t)\|_{p,s_1} \leq C(t \wedge 1)^{-\mathfrak{n}\left(\frac{1}{s_1} - \frac{1}{p}\right)+\alpha-1} \leq C(t \wedge 1)^{-\beta+\alpha-1}$$

for all $t > 0$. We divide the proof into two cases according to whether $t \leq 1$ or $t > 1$. In the first case ($t \wedge 1 \equiv t$), we have by the Hölder inequality with exponents (s, s_2),

$$\begin{aligned} |g(t)| &= t^\delta \int_0^t \|P_\alpha(t-\tau)\|_{p,s_1} \tau^{-\theta} r(\tau)\, d\tau \\ &\leq Ct^\delta \int_0^t (t-\tau)^{-(\beta+1-\alpha)} \tau^{-\theta} r(\tau)\, d\tau \\ &\leq Ct^\delta \left(\int_0^t (t-\tau)^{-(\beta+1-\alpha)s} \tau^{-\theta s} d\tau \right)^{1/s} p_{s_2}(r). \end{aligned} \tag{A.0.4}$$

Now, by a basic change of variable $s = \tau/t$, we have

$$\begin{aligned} &\left(\int_0^t (t-\tau)^{-(\beta+1-\alpha)s} \tau^{-\theta s} d\tau \right)^{1/s} \\ &= t^{1/s-(\beta+1-\alpha)-\theta} \left(\int_0^1 (1-s)^{-(\beta+1-\alpha)s} s^{-\theta s} ds \right)^{1/s} \\ &\leq Ct^{1/s-(\beta+1-\alpha)-\theta}, \end{aligned} \tag{A.0.5}$$

where the last integral on the right-hand side converges provided that $\theta s < 1$ ($\Leftrightarrow \theta < 1 - \frac{1}{s_2}$) and $\beta + 1 - \alpha < 1/s = 1 - 1/s_2$. Note that the first condition coincides with the first assumption of (A.0.2). Therefore, from (A.0.4) we deduce

$$|g(t)| \leq Ct^{1/s-(\beta+1-\alpha)-\theta+\delta} p_{s_2}(r) \leq Ct^{\varepsilon} p_{s_2}(r), \text{ for } t \leq 1,$$

where we have noticed once again the second of (A.0.3). The proof of the case $t > 1$ can be reduced to the first case by choosing $k \in \mathbb{N}$ such that $k < t \leq k+1$ (i.e., $t \wedge 1 \equiv 1$). Indeed, we can separate the integral in the definition of $g(t)$ into a sum

$$\int_0^t = \int_0^1 + \int_1^2 + \ldots + \int_{k-2}^{k-1} + \int_{k-1}^{t-1} + \int_{t-1}^t,$$

and then apply the Hölder inequality to each summand separately. Observe that this decomposition allows the restriction $t_2 - t_1 \leq 1$ in the definition of $p_{s_2}(r)$. By a similar reasoning to (A.0.5), we get

$$\begin{aligned}
|g(t)| &\leq Cp_{s_2}(r) \sum \left(\int (t-\tau)^{-(\beta+1-\alpha)s} \tau^{-\theta s} d\tau \right)^{1/s} \\
&\leq Cp_{s_2}(r) \left(\left(\int_0^1 \tau^{-\theta s} d\tau \right)^{1/s} + 1 + \ldots + 1 + \int_{t-1}^t (t-\tau)^{-(\beta+1-\alpha)s} d\tau \right) \\
&\leq Cp_{s_2}(r)(k+1) \\
&\leq 2(Ct)\, p_{s_2}(r),
\end{aligned}$$

Here we have majorized all the intermediate summands by the value one since on each of the corresponding intervals the integrands are bounded above by 1. This completes the proof of the lemma. □

The ultracontractivity property of the operator S_α (see (2.2.12)) allows us to also deduce the following lemma.

Lemma A.0.2 *Let $p, q \in [1, \infty]$ such that $p < q$ and set $\overline{\delta} := \mathfrak{n}/p - \mathfrak{n}/q \in (0, \alpha)$. Given a subset $\Pi \subset L^p(\mathcal{X})$, assume that the set*

$$\kappa(\Pi) := \left\{ u \|u\|_{L^p(\mathcal{X})}^{-1} : u \in \Pi,\ u \neq 0 \right\}$$

is precompact in $L^p(\mathcal{X})$. Then there exists a continuous and nondecreasing function $g : [0, \infty) \to [0, 1]$, depending only on p, q, δ and the set Π such that the following assertions hold.

(a) For all $t > 0$ and $u \in \Pi$,

$$\|S_\alpha(t) u\|_{L^q(\mathcal{X})} \leq Cg(t)(t \wedge 1)^{-\overline{\delta}} \|u\|_{L^p(\mathcal{X})}. \tag{A.0.6}$$

(b) We have $\lim_{t\to 0^+} g(t) = 0$. *The function* $w = w(t)$ *defined by*

$$(w(t))^{-\overline{\delta}} = g(t)(t \wedge 1)^{-\overline{\delta}}$$

has the properties

$$\lim_{t\to 0^+} w(t) = 0 \text{ and } (t \wedge 1) \leq w(t) \leq (t \wedge 1)^{\frac{1}{2}}.$$

Proof The proof of this statement is similar to that of [44, Lemma 4] but requires some nontrivial modifications since S_α is *not* a semigroup for $\alpha \in (0,1)$. By assumption, the set $\kappa(\Pi)$ is precompact in $L^p(\mathcal{X})$. Since $p < \infty$ and $\mathcal{X}$ is a relatively compact (Hausdorff) space, $L^q(\mathcal{X}) \subset L^p(\mathcal{X})$ is a dense subset of $L^p(\mathcal{X})$. Hence, for any $\varepsilon > 0$, there exists a finite set $\{v_1, \ldots, v_M\} \subset L^q(\mathcal{X}) \setminus \{0\}$ such that

$$\min_{1\leq m\leq M} \left\| u \|u\|^{-1}_{L^p(\mathcal{X})} - v_m \right\|_{L^p(\mathcal{X})} \leq \varepsilon, \text{ for all } u \in \Pi,\ u \neq 0. \tag{A.0.7}$$

Define now the function $h : (t,u) \in (0,\infty) \times \Pi \setminus \{0\} \to h(t,u) \in [0,1]$, by

$$h(t,u) := \|S_\alpha(t) u\|_{L^q(\mathcal{X})} C_0^{-1} t^{\overline{\delta}} \|u\|^{-1}_{L^p(\mathcal{X})},$$

where $C_0 (= C) > 0$ is the constant from the ultracontractivity estimate (2.2.4) for the operator S_α. We observe that for all $v_m \in L^q(\mathcal{X}) \setminus \{0\}$, $m = 1, \ldots, M$, we have owing to the fact that $\|S_\alpha(t)\|_{q,q} \leq 1$,

$$\begin{aligned} 0 \leq h(t, v_m) &= \|S_\alpha(t) v_m\|_{L^q(\mathcal{X})} C_0^{-1} t^{\overline{\delta}} \|v_m\|^{-1}_{L^p(\mathcal{X})} \\ &\leq C_0^{-1} t^{\overline{\delta}} \|v_m\|^{-1}_{L^p(\mathcal{X})} \|v_m\|_{L^q(\mathcal{X})}. \end{aligned}$$

Therefore, since $\overline{\delta} > 0$ it holds

$$\lim_{t\to 0^+} h(t, v_m) = 0, \text{ for all } v_m \in L^q(\mathcal{X}) \setminus \{0\},\quad m = 1, \ldots, M. \tag{A.0.8}$$

Clearly, by (A.0.7) we also have $\|v_m\|_{L^p(\mathcal{X})} \leq (1+\varepsilon)$. Next, we estimate for a suitably chosen $m \in \{1, .., M\}$,

$$\begin{aligned} & C_0^{-1} t^{\overline{\delta}} \left\| S_\alpha(t) u \|u\|^{-1}_{L^p(\mathcal{X})} \right\|_{L^q(\mathcal{X})} \\ & \leq C_0^{-1} t^{\overline{\delta}} \left\| S_\alpha(t) \left(u \|u\|^{-1}_{L^p(\mathcal{X})} - v_m \right) \right\|_{L^q(\mathcal{X})} + C_0^{-1} t^{\overline{\delta}} \|S_\alpha(t) v_m\|_{L^q(\mathcal{X})} \\ & \leq \left\| u \|u\|^{-1}_{L^p(\mathcal{X})} - v_m \right\|_{L^p(\mathcal{X})} + C_0^{-1} t^{\overline{\delta}} \|S_\alpha(t) v_m\|_{L^q(\mathcal{X})} \end{aligned} \tag{A.0.9}$$

$$\le \varepsilon + C_0^{-1} t^{\overline{\delta}} \left\| S_\alpha(t) v_m \right\|_{L^q(\mathcal{X})}$$
$$\le \varepsilon + (1+\varepsilon) h(t, v_m),$$

where in the first term we have applied the ultracontractivity estimate for S_α. Also, in the estimate we have recalled (A.0.7) and the fact that, by definition,

$$h(t, v_m) \left\| v_m \right\|_{L^p(\mathcal{X})} = \left\| S_\alpha(t) v_m \right\|_{L^q(\mathcal{X})} C_0^{-1} t^{\overline{\delta}}.$$

Consequently, (A.0.9) then yields

$$h(t, u) \le \varepsilon + (1+\varepsilon) h(t, v_m), \text{ for all } u \in \Pi,\ u \neq 0. \tag{A.0.10}$$

Define now a new function $\overline{h}(t) = \sup\{h(s,u) : s \in [0,t], u \in \Pi \backslash \{0\}\}$. By virtue of (A.0.8) and estimate (A.0.10), we have that $0 \le \overline{h} \le 1$ and $\lim_{t \to 0^+} \overline{h}(t) = 0$. The definitions of h and $\overline{h}$ allow one to get, for $t > 0$,

$$\left\| S_\alpha(t) u \right\|_{L^q(\mathcal{X})} \le C_0 \overline{h}(t) t^{-\overline{\delta}} \left\| u \right\|_{L^p(\mathcal{X})}, \text{ for all } u \in \Pi \backslash \{0\}.$$

Setting $g(t) = \sup\left\{\overline{h}(t), (t \wedge 1)^{\overline{\delta}/2}\right\}$, all the assertions (a)–(b) of the lemma are fulfilled by the operator S_α. The proof is complete. □

Lemma A.0.3 *Consider the following cases:*

(a) Let $p_0, \gamma \in [1, \infty)$, $q_1 \in [1,\infty] \cap (\beta_A, \infty]$, $q_2 \in (1/\alpha, \infty]$ *satisfy*

$$\frac{\mathfrak{n}}{q_1} + \frac{1}{q_2} + (\gamma - 1)\frac{\mathfrak{n}}{p_0} < \alpha. \tag{A.0.11}$$

(b) Let $p_0, \gamma \in (1, \infty)$, $q_1 \in [1,\infty] \cap (\beta_A, \infty]$, $q_2 \in (1/\alpha, \infty]$ *satisfy*

$$\frac{\mathfrak{n}}{q_1} + \frac{1}{q_2} + (\gamma - 1)\frac{\mathfrak{n}}{p_0} = \alpha.$$

Then there exist $\varepsilon \in (0, \alpha)$, $k \in \mathbb{N}$ *and finite sequences* $\{p_i\}$, $\{\delta_i\}$ *such that* $\delta_i \in (0, \alpha)$ *and* $p_0 < p_1 < \ldots < p_k = \infty$, *for* $i = 1, .., k$. *In addition, the following are satisfied:*

$$\begin{cases} \frac{1}{q_2} + \frac{\mathfrak{n}}{q_1} + (\gamma - 1)(\delta_i + \frac{\mathfrak{n}}{p_i}) + \varepsilon < \alpha, & \text{for } i = 1, .., k;\ i \neq 1 \text{ in case (b).} \\ \dfrac{1}{q_1} + \dfrac{\gamma}{p_i} \le 1, & \text{for } i = 1, .., k. \\ \dfrac{1}{q_2} + \gamma \delta_i < 1, & \text{for } i = 1, .., k. \\ \dfrac{\mathfrak{n}}{p_{i-1}} - \dfrac{\mathfrak{n}}{p_i} =: \delta_i, & \text{for } i = 1, .., k. \end{cases} \tag{A.0.12}$$

Proof The proof is similar to that of [44, Lemma 12]. We include the details for the proof of (a) for the sake of completeness. The case (b) is similar. The crucial point is to choose $p_1 \in [1, \infty]$ such that the first and second of (A.0.12) hold with $\delta_1 \in (0, \alpha)$ given by the fourth condition for $i = 1$. These conditions can be written more clearly as follows:

$$\frac{\gamma \mathfrak{n}}{p_0} > \frac{\gamma \mathfrak{n}}{p_1}, \quad \frac{\gamma \mathfrak{n}}{p_1} \leq \mathfrak{n} - \frac{\mathfrak{n}}{q_1} \tag{A.0.13}$$

and

$$\frac{1}{q_2} + \frac{\gamma \mathfrak{n}}{p_0} - \alpha < \frac{\gamma \mathfrak{n}}{p_1}. \tag{A.0.14}$$

It turns out that there exists a value $p_1 \in [1, \infty]$ such that (A.0.13)–(A.0.14) are satisfied if and only if the following hold:

$$\frac{1}{p_0} > 0, \frac{1}{q_1} \leq 1, \ \frac{1}{q_2} < \alpha < 1 \text{ and } \frac{1}{q_2} + \frac{\mathfrak{n}}{q_1} + (\gamma - 1)\frac{\mathfrak{n}}{p_0} < \alpha + \mathfrak{n}\left(1 - \frac{1}{p_0}\right). \tag{A.0.15}$$

But the assumed condition (A.0.11) (which also coincides with the first of (A.0.12) for $i = 1$) automatically implies the last of (A.0.15); the other conditions of (A.0.15) are also satisfied since $q_1 \in [1, \infty] \cap (\beta_A, \infty]$, $q_2 \in (1/\alpha, \infty]$. Thus, there exist p_1 and δ_1 satisfying (A.0.12) for $i = 1$. Next choose $p_2 < p_3 < \ldots < p_k = \infty$ and $\delta_i = \mathfrak{n}/p_{i-1} - \mathfrak{n}/p_i$, for $i = 2, \ldots, k$, such that $\delta_i \leq \delta_1$ for $i = 2, \ldots, k$. Then all the assertions of the lemma are obviously satisfied. □

We have used the following simple estimates repeatedly in Chap. 3.

Lemma A.0.4 *The following assertions hold.*

(i) Let $\varepsilon \in (0, 1]$ and $0 \leq t \leq t + h \leq T$. Then there exists $q > 1/\varepsilon \geq 1$ such that

$$(t+h)^\varepsilon - t^\varepsilon \leq \frac{\varepsilon\,(q-1)}{\varepsilon q - 1} T^{\frac{\varepsilon q - 1}{q-1}} h^{1/q}.$$

(ii) For $a \geq b \geq 0$ and $q \geq 1$, the following inequality holds:

$$(a-b)^q \leq a^q - b^q.$$

Proof

(i) The case $\varepsilon = 1$ is obvious. Let $p > 1$ such that $p\,(1 - 1/q) = 1$. We estimate

$$(t+h)^\varepsilon - t^\varepsilon = \varepsilon \int_t^{t+h} s^{\varepsilon-1} ds \leq \varepsilon \left(\int_t^{t+h} s^{(\varepsilon-1)p} ds \right)^{1/p} h^{1/q}$$

$$\leq \frac{\varepsilon}{(\varepsilon-1)\,p+1}\,(t+h)^{(\varepsilon-1)p+1}\,h^{1/q}$$
$$\leq \frac{\varepsilon\,(q-1)}{\varepsilon q-1}\,T^{\frac{\varepsilon q-1}{q-1}}\,h^{1/q},$$

since $(\varepsilon-1)\,p > -1$ owing to $q\varepsilon > 1$. To prove the second claim (ii), we first notice that equality holds when $b = 0$ and $a \geq 0$ or $q = 1$. Thus we may assume that $b > 0$, $q > 1$ and denote by $x = a/b \geq 1$. The inequality we need to prove is equivalent to $(x-1)^q \leq x^q - 1$, for all $x \geq 1$. Set then $h\,(x) = x^q - 1 - (x-1)^q$, and notice that $h'\,(x) = qx^{q-1} - q\,(x-1)^{q-1} \geq 0$, due to $q > 1$ and $x \geq 1$. Hence, $h\,(x) \geq h\,(1) = 0$ and the claim follows. □

Lemma A.0.5 *Let* $0 < \alpha \leq 1$. *Let* $p_0 \in [1,\infty]$ *be arbitrary and* $q_1 \in (\beta_A,\infty] \cap [1,\infty]$, $q_2 \in (1/\alpha,\infty]$, $r_1, r_2 \in (0,\infty]$, $\gamma \in [1,\infty)$, $b \in [0,1]$ *such that*

$$\frac{\mathfrak{n}}{q_1}+\frac{1}{q_2}+\gamma\,(1-b)\left(\frac{\mathfrak{n}}{r_1}+\frac{1}{r_2}\right) < \alpha,\quad \frac{1}{q_1}+\frac{\gamma\,(1-b)}{r_1} \leq 1 \text{ and } \gamma b \leq 1.$$

Then there exist $\varepsilon \in (0,\alpha)$, $k \in \mathbb{N}$ *and finite sequences* $\{p_i\}$, $\{\delta_i\}$ *such that* $\delta_i \in (0,\alpha)$ *and* $p_0 < p_1 < \ldots < p_k = \infty$, *for* $i = 1,..,k$. *In addition, the following are satisfied:*

$$\begin{cases} \dfrac{1}{q_1}+\gamma\dfrac{1-b}{r_1}+\gamma\dfrac{b}{p_i} \leq 1, & \text{for } i = 1,..,k;\\ \dfrac{1}{q_2}+\gamma\dfrac{1-b}{r_2}+\gamma b\delta_i < \alpha-\varepsilon, & \text{for } i = 1,..,k.\\ \delta_i = \dfrac{\mathfrak{n}}{p_{i-1}}-\dfrac{\mathfrak{n}}{p_i}, & \text{for } i = 1,..,k. \end{cases}$$

Proof The proof follows closely that of [44, Lemma 16] with some minor (inessential) modifications; we leave it to the interested reader. □

The following basic "feedback" inequality is taken from [44, Lemma 18].

Lemma A.0.6 *Let* $y, z_0, z_1 \in [0,\infty)$ *and* $\sigma \in (0,1)$ *be such that* $y \leq z_0 + z_1 y^\sigma$. *Then*

$$y \leq \frac{z_0}{1-\sigma} + z_1^{\frac{1}{1-\sigma}}.$$

We next state a simple comparison principle for some ordinary differential equation associated with ∂_t^α.

Lemma A.0.7 *Let* $y_i \in C\,[0,T]$ *such that* $g_{1-\alpha} * y_i \in C^1\,(0,T)$, *for* $i = 1, 2$ *and let* $a \in \mathbb{R}$, $f \in L^1\,[0,T]$. *Assume that* y_i *satisfy the following inequalities, for* *almost all* $t \in (0, T]$,

$$\begin{cases} \partial_t^\alpha y_1(t) + a y_1(t) \le f(t), \\ \partial_t^\alpha y_2(t) + a y_2(t) \ge f(t), \end{cases}$$

and $y_1(0) \le y_2(0)$. *Then* $y_1(t) \le y_2(t)$ *on* $[0, T]$.

Proof Let us set $u := y_1 - y_2$, and subtract the second equation from the first equation. We obtain the inequality

$$\partial_t^\alpha u(t) + au(t) \le 0, \; u(0) = y_1(0) - y_2(0) \le 0.$$

Observe that the unique solution of $\partial_t^\alpha z(t) + az(t) = 0$, with $z(0) = u_0$, is given by $z(t) = E_{\alpha,1}(-at^\alpha) z_0$ (see, for instance, [5, Corollary 2.39]). By the comparison principle for linear fractional differential equations (see [19]), we then deduce that $u(t) \le z(t)$, $t \in [0, T]$. Since $z(0) = z_0 \le 0$, it follows that $u(t) \le z(t) \le 0$, which is the desired claim. □

We now state an important inequality that allows one to deduce uniform bounds with respect to time and with respect to the parameter $\alpha \to 1^-$.

Lemma A.0.8 *Let* $T \in (0, \infty]$ *be given. Let* $y \in C[0, T]$ *such that* $g_{1-\alpha} * y \in C^1(0, T)$ *and let* $a > 0$, $0 \le f \in C[0, T)$ *such that* $\sup_{t \in [0,T)} f(t) = M > 0$. *Suppose that (a nonnegative) y satisfies the inequality*

$$\partial_t^\alpha y(t) + ay(t) \le f(t), \; a.e. \text{ on } (0, T),$$

such that $y(0) = y_0 \ge 0$. *Then*

$$\sup_{t \in [0,T)} [y(t)] \le C \max\left\{ y_0, \frac{M}{a^{1/\alpha}} \right\}, \tag{A.0.16}$$

for some $C = C(\alpha) > 0$ *(independent of* t, T, y_0 *and* y*), which is bounded as* $\alpha \to 1^-$. *Furthermore, if* $T = \infty$, *we have*

$$\limsup_{t \to \infty} [y(t)] \le C \frac{M}{a^{1/\alpha}}, \; \alpha \in (0, 1). \tag{A.0.17}$$

Proof By assumption, f is bounded on $[0, T)$. Let $z \ge 0$ be the corresponding (unique) solution for the problem

$$\begin{cases} \partial_t^\alpha z(t) + az(t) = f(t), \; t \in (0, T), \\ z(0) = y(0) = y_0 \ge 0. \end{cases}$$

This solution is given by

$$z(t) = y_0 E_{\alpha,1}\left(-at^\alpha\right) + \int_0^t (t-s)^{\alpha-1} E_{\alpha,\alpha}\left(-a(t-s)^\alpha\right) f(s)\, ds, \qquad \text{(A.0.18)}$$

see, for instance, [5, Chapter 3]. We know that for any $x \le 0$, $\beta_1 \in (0,2)$ and $\beta_2 > 0$,

$$E_{\beta_1,\beta_2}(x) \le \frac{C}{1+|x|},$$

for some $C = C(\beta_1, \beta_2) > 0$ (see [41]). Furthermore, we have the following formula (cf. [34, Chapter 2]) for any $\xi > 0$ and $b \in \mathbb{R}$,

$$\int_0^\xi s^{\alpha-1} E_{\alpha,\alpha}\left(bs^\alpha\right) ds = b\xi^\alpha E_{\alpha,\alpha+1}\left(b\xi^\alpha\right). \qquad \text{(A.0.19)}$$

It follows from (A.0.18) that

$$\begin{aligned} z(t) &\le y_0 \frac{C(\alpha)}{1+at^\alpha} + M \int_0^t (t-s)^{\alpha-1} E_{\alpha,\alpha}\left(-a(t-s)^\alpha\right) ds \qquad \text{(A.0.20)} \\ &\le y_0 \frac{C(\alpha)}{1+at^\alpha} + \frac{M}{a^{1/\alpha}} \int_0^{a^{1/\alpha}t} x^{\alpha-1} E_{\alpha,\alpha}\left(-x^\alpha\right) dx \\ &= y_0 \frac{C(\alpha)}{1+at^\alpha} + \frac{M}{a^{1/\alpha}} \left(at^\alpha E_{\alpha,\alpha+1}\left(-at^\alpha\right)\right) \\ &\le y_0 \frac{C(\alpha)}{1+at^\alpha} + Ma^{1-1/\alpha} C(\alpha) \frac{t^\alpha}{1+at^\alpha} \\ &\le 2C(\alpha) \max\left\{y_0, Ma^{-1/\alpha}\right\}, \end{aligned}$$

for all $t \in (0,T)$. The constant $C(\alpha) > 0$ is bounded as $\alpha \to 1$. Finally, estimate (A.0.20) together with the comparison principle (see Lemma A.0.7), which yields that $y(t) \le z(t)$ on $[0,T)$, gives (A.0.16). □

We recall the following version of Grönwall lemma [26, Lemma 7.1.1] (cf. also [5, Theorem 2.19] for a proof).

Lemma A.0.9 *Given $b \ge 0$ and $0 \le l, \omega \in L^1_{loc}(\mathbb{R}_+)$, satisfying*

$$\omega(t) \le l(t) + \frac{b}{\Gamma(\alpha)} \int_0^t (t-s)^{\alpha-1} \omega(s)\, ds,$$

it holds

$$\omega(t) \leq l(t) + b \int_0^t (t-s)^{\alpha-1} E_{\alpha,\alpha}\left(b(t-s)^{\alpha}\right) l(s)\, ds,$$

for all $t \in [0, T]$.

Appendix B
Integration by Parts Formula for the Regional Fractional Laplacian

The main objective of this appendix is to give an integration by parts formula for the regional fractional Laplace which can be used to define the fractional Neumann and/or Robin boundary conditions. To do this we first introduce a notion of fractional normal derivative that we have mentioned in the previous chapters. We assume first that $\Omega \subset \mathbb{R}^N$ is a bounded open set with boundary of class $C^{1,1}$. We will also use the following notations:

$$\rho(x) = \text{dist}(x, \partial\Omega) = \inf\{|y - x| : y \in \partial\Omega\}, \; \forall\, x \in \Omega,$$
$$\Omega_\delta = \{x \in \Omega : 0 < \rho(x) < \delta\}, \; \delta > 0 \text{ is a real number},$$
$$\nu(z) = \text{ the outer normal vector of } \partial\Omega \text{ at the point } z \in \partial\Omega.$$

The following definition is taken from [20, Definition 2.1] (see also [22, Definition 7.1] for the one-dimensional case).

Definition B.0.1 For $u \in C^1(\Omega)$, $z \in \partial\Omega$ and $0 \leq \alpha < 2$, we define the boundary operator $\mathcal{N}^\alpha$ by

$$\mathcal{N}^\alpha u(z) = \lim_{t \downarrow 0} \frac{du(z + \boldsymbol{\nu}(z)t)}{dt} t^\alpha, \tag{B.0.1}$$

whenever the limit exists.

Remark B.0.2 Let $0 \leq \alpha < 2$ and let $\mathcal{N}^\alpha$ be the boundary operator defined in (B.0.1).

(a) It is easy to see that if $u \in C^1(\Omega) \cap C(\overline{\Omega})$ then for every $z \in \partial\Omega$,

$$\mathcal{N}^\alpha u(z) = \lim_{t \downarrow 0} \frac{u(z + t\boldsymbol{\nu}(z)) - u(z)}{t^{2s-1}}.$$

C. G. Gal, M. Warma, *Fractional-in-Time Semilinear Parabolic Equations and Applications*, Mathématiques et Applications 84,
https://doi.org/10.1007/978-3-030-45043-4

(b) If $\alpha = 0$, then $\mathcal{N}^0 u(z) = \nabla u \cdot \nu(z) = \frac{\partial u(z)}{\partial \nu}$ for every $u \in C^1(\overline{\Omega})$ and $z \in \partial\Omega$.
(c) If $0 < \alpha < 2$, then $\mathcal{N}^\alpha u(z) = 0$ for every $u \in C^1(\overline{\Omega})$ and $z \in \partial\Omega$.

For $\beta > 0$ we define the space

$$C^2_\beta(\overline{\Omega}) = \Big\{u : \ u(x) = f(x)\rho(x)^{\beta-1} + g(x), \ \forall\, x \in \Omega, \ \text{ for some } \ f, g \in C^2(\overline{\Omega})\Big\}.$$

When $\beta > 1$, we always assume that $u \in C^2_\beta(\overline{\Omega})$ is defined on $\overline{\Omega}$ by continuous extension. The following explicit representation of the operator $\mathcal{N}^\alpha$ is taken from [54, Lemma 5.3].

Lemma B.0.3 *Let $1 < \beta \leq 2$ and $u \in C^2_\beta(\overline{\Omega})$. Then the following assertions hold.*

(a) If $\beta \in (1, 2)$, then for $z \in \partial\Omega$,

$$\mathcal{N}^{2-\beta} u(z) = (1-\beta) \lim_{\Omega \ni x \to z} \frac{u(x) - u(z)}{\rho(x)^{\beta-1}}. \tag{B.0.2}$$

(b) If $\beta = 2$, then for $z \in \partial\Omega$,

$$\mathcal{N}^0 u(z) = -\lim_{\Omega \ni x \to z} \frac{u(x) - u(z)}{\rho(x)}. \tag{B.0.3}$$

Next, for $\frac{1}{2} < s, 1$, we let

$$C_s := \frac{C_{1,s}}{2s(2s-1)} \int_0^\infty \frac{|\tau - 1|^{1-2s} - (\tau \vee 1)^{1-2s}}{\tau^{2-2s}} d\tau.$$

One can show that $\lim_{s\uparrow 1} C_s = 1$. For more details on this topic we refer to [15, 53–55] and their references.

We have the following fractional Green type formula for the regional fractional Laplace operator.

Theorem B.0.4 *Let $\frac{1}{2} < s < 1$. Then, for every $u \in C^2_{2s}(\overline{\Omega})$ and $v \in W^{s,2}(\Omega)$ we have that*

$$\int_\Omega v(x)(-\Delta)^s_\Omega u(x)dx = \frac{C_{N,s}}{2} \int_\Omega \int_\Omega \frac{(u(x) - u(y))(v(x) - v(y))}{|x-y|^{N+2s}} dxdy \tag{B.0.4}$$

$$- C_s \int_{\partial\Omega} v\mathcal{N}^{2-2s} u d\sigma.$$

We mention that the identity (B.0.4) has been first obtained in [20, Theorem 3.3] under the assumption that v also belongs to $C^2_{2s}(\overline{\Omega})$. Its validity for every $v \in W^{s,2}(\Omega)$ has been proved in [54, Theorem 5.7] by using a density argument (see also [55] for a more general operator).

Definition B.0.5 For $\frac{1}{2} < s < 1$ and $u \in C^2_{2s}(\overline{\Omega})$, we call the function $C_s\mathcal{N}^{2-2s}u$ ***the strong fractional normal derivative*** of u in direction of the outer normal vector.

We make some comments about the fractional normal derivative introduced above.

Remark B.0.6 It is worth to observe the following facts.

(a) The nonlocal normal derivative, has been introduced in [8, 23] (see also [7]) for functions u defined on $\mathbb{R}^N$. More precisely, recall that for $0 < \alpha < 1$ and $u \in \mathcal{L}^1(\mathbb{R}^N)$, the non-local normal derivative is defined by

$$\mathcal{N}_\alpha u(x) = C_{N,s}\int_\Omega \frac{u(x)-u(y)}{|x-y|^{N+2\alpha}}dy, \quad x \in \mathbb{R}^N\backslash\overline{\Omega}, \tag{B.0.5}$$

provided that the integral exists. The definition of $\mathcal{N}_\alpha u$ in (B.0.5) requires that the function is defined on all $\mathbb{R}^N$. This is different from the fractional Normal derivative $\mathcal{N}^\alpha u$ given in (B.0.1) where the function u is defined only on Ω. Starting with a function defined only on Ω, it seems impossible to deal with $\mathcal{N}_\alpha u$. For example if $u \in W^{s,2}(\Omega)$ and letting $\tilde{u} \in W^{s,2}(\mathbb{R}^N)$ be an extension to all $\mathbb{R}^N$, then the relation (B.0.5) can make sense but the definition cannot be independent of the extension, except in the case where there is only one such possible extension. This shows that the expression $\mathcal{N}_\alpha u$ cannot be used in the case of the fractional Laplace operator where one considers functions defined a priori only on Ω. It has been shown in [7, Proposition 5.1] (see also [8, 23]) that if $\Omega \subset \mathbb{R}^N$ is a bounded domain with Lipschitz continuous boundary $\partial\Omega$, then for every $u, v \in C_0^2(\mathbb{R}^N)$,

$$\lim_{\alpha\uparrow 1}\int_{\mathbb{R}^N\backslash\Omega} v\mathcal{N}_\alpha u dx = \int_{\partial\Omega}\frac{\partial u}{\partial \nu}v d\sigma.$$

(b) As we have seen in Remark B.0.2, the fractional normal derivative $\mathcal{N}^\alpha u$ is continuous with respect to α, so that for every $u \in C^2_2(\overline{\Omega}) = C^1(\overline{\Omega})$ we have that $\mathcal{N}^1 u = \frac{\partial u}{\partial \nu}$, i.e., the classical normal derivative of u in direction of the outer normal vector $\boldsymbol{\nu}$.

Next, we introduce a weak formulation on non-smooth domains of a fractional normal derivative.

Definition B.0.7 Let $\frac{1}{2} < s < 1$ and $\Omega \subset \mathbb{R}^N$ a bounded domain with Lipschitz continuous boundary $\partial\Omega$.

(a) Let $u \in W^{s,2}(\Omega)$. We say that $(-\Delta)^s_\Omega u \in L^2(\Omega)$ if there exists $w \in L^2(\Omega)$ such that

$$\frac{C_{N,s}}{2}\int_\Omega\int_\Omega \frac{(v(x)-v(y))(u(x)-u(y))}{|x-y|^{N+2s}}dxdy = \int_\Omega wv dx$$

for all $v \in W_0^{s,2}(\Omega)$. In that case we write $(-\Delta)^s_\Omega u = w$.

(b) Let $u \in W^{s,2}(\Omega)$ such that $(-\Delta)^s_\Omega u \in L^2(\Omega)$. We say that u has a fractional normal derivative in $L^2(\partial\Omega)$ if there exists $g \in L^2(\partial\Omega)$ such that

$$\int_\Omega v(-\Delta)^s_\Omega u dx = \frac{C_{N,s}}{2}\int_\Omega\int_\Omega \frac{(v(x)-v(y))(u(x)-u(y))}{|x-y|^{N+2s}}dxdy - \int_{\partial\Omega} gv d\sigma \tag{B.0.6}$$

for all $v \in W^{s,2}(\Omega)$. In that case, the function g is uniquely determined by (B.0.6), we write $C_s\mathcal{N}^{2-2s}u = g$ and call g the **weak fractional normal derivative** of u.

Using Definition B.0.7 and an approximation argument we get the following more general Green's type formula for the regional fractional Laplace operator.

Theorem B.0.8 *Let $\frac{1}{2} < s < 1$ and let $\Omega \subset \mathbb{R}^N$ be a bounded open set with Lipschitz continuous boundary. Then for all $v \in W^{s,2}(\Omega)$ the identity*

$$\int_\Omega v(-\Delta)^s_\Omega u\, dx = \frac{C_{N,s}}{2}\int_\Omega\int_\Omega \frac{(v(x)-v(y))(u(x)-u(y))}{|x-y|^{N+2s}}dxdy \tag{B.0.7}$$
$$- C_s\int_{\partial\Omega} v\mathcal{N}^{2-2s}u d\sigma,$$

holds, whenever $u \in W^{s,2}(\Omega)$, $(-\Delta)^s_\Omega u \in L^2(\Omega)$ and $\mathcal{N}^{2-2s}u$ exists in $L^2(\partial\Omega)$.

We mention that if Ω is a bounded open set of class $C^{1,1}$ and $u \in C^2_{2s}(\overline{\Omega})$, then by [20, 54], $\mathcal{N}^{2-2s}u \in L^2(\partial\Omega)$, $(-\Delta)^s_\Omega u \in L^2(\Omega)$ and in that case weak and strong fractional normal derivatives of u coincide in the sense that they are equal everywhere on $\partial\Omega$.

Appendix C
A Zoo of Fractional Kinetic Equations

Fractional kinetic equations involving diffusion and/or diffusion-advection provide a useful approach for the description of transport dynamics in complex systems which are governed by anomalous diffusion and non-exponential relaxation patterns. Such fractional equations are usually derived asymptotically from basic random walk models [36]; among them we quote the fractional Brownian motion, the continuous time random walk, the Lévy flight, the Schneider-Grey Brownian motion and, more generally, random walk models based on evolution equations of single and distributed fractional order in time and/or in space [9, 25, 33, 47, 51]. Indeed, in mathematical physics it is often more convenient to have a deterministic equation for the probability density function of a process, given as the analogue of the classical heat equation, to be solved under given initial and boundary conditions.

C.1 Fractional Equation with Nonlocality in Space

We present next a diffusion equation that is generated by Lévy statistics, and can be also derived asymptotically from a simple exclusion process with long-range random jumps [29]. From the probabilistic point of view, the best way to understand Lévy statistics when compared to Brownian statistics is the random walk formalism. In the latter, particles make *only* small steps with finite probability such that the interaction between close neighbors is always short-ranged while in the former, particles are also allowed to take "arbitrarily" large steps (up to the system size) with a small finite probability for each such step, and so the interaction between particles is long-ranged. Physical phenomena that exhibit deviations from normal diffusion is usually dubbed as anomalous diffusion and seems to be inherent in dynamical systems far from equilibrium [39, 49, 52]. Let $\mathcal{K}: \mathbb{R}^N \to [0,\infty)$ be an

C. G. Gal, M. Warma, *Fractional-in-Time Semilinear Parabolic Equations and Applications*, Mathématiques et Applications 84,
https://doi.org/10.1007/978-3-030-45043-4

even function such that

$$\sum_{k\in\mathbb{Z}^N} \mathcal{K}(k) = 1. \tag{C.1.1}$$

Given a small $h > 0$, we consider a random walk on the lattice $h\mathbb{Z}^N$. We suppose that at any unit time τ (which may depend on h) a particle jumps from any point of $h\mathbb{Z}^N$ to any other point. The probability for which a particle jumps from a point $hk \in h\mathbb{Z}^N$ to the point $h\tilde{k}$ is taken to be $\mathcal{K}(k-\tilde{k}) = \mathcal{K}(\tilde{k}-k)$. Note that, differently from the standard random walk, in this process the particle may experience arbitrarily long jumps, though with small probability. Let $u(x,t)$ be the probability that our particle lies at $x \in h\mathbb{Z}^N$ at time $t \in \tau\mathbb{Z}$. Then $u(x, t+\tau)$ is the sum of all the probabilities of the possible positions $x + hk$ at time t weighted by the probability of jumping from $x + hk$ to x. That is,

$$u(x, t+\tau) = \sum_{k\in\mathbb{Z}^N} \mathcal{K}(k)u(x+hk, t).$$

Using (C.1.1) we have the evolution law:

$$u(x, t+\tau) - u\,(x, t) = \sum_{k\in\mathbb{Z}^N} \mathcal{K}(k)\,[u(x+hk, t) - u(x, t)]\,. \tag{C.1.2}$$

In particular, in the case when $\tau = h^{2s}$ and $\mathcal{K}$ is homogeneous (i.e., $\mathcal{K}(y) = |y|^{-(N+2s)}$ for $y \neq 0$, $\mathcal{K}(0) = 0$, and $0 < s < 1$), (C.1.1) holds and $\mathcal{K}(k)/\tau = h^N\mathcal{K}(hk)$. Therefore, we can rewrite (C.1.2) as follows:

$$\frac{u(x, t+\tau) - u(x, t)}{\tau} = h^N \sum_{k\in\mathbb{Z}^N} \mathcal{K}(hk)\,[u(x+hk, t) - u(x, t)]\,. \tag{C.1.3}$$

Notice that the term on the right-hand side of (C.1.3) is just the approximating Riemann sum of

$$\int_{\mathbb{R}^N} \mathcal{K}(y)\,[u(x+y, t) - u(x, t)]\,dy.$$

Thus letting $\tau = h^{2s} \to 0^+$ in (C.1.3), we obtain

$$\partial_t u(x, t) = \int_{\mathbb{R}^N} \frac{u(x+y, t) - u(x, t)}{|y|^{N+2s}}dy. \tag{C.1.4}$$

The integral on the right-hand side of (C.1.4) has a singularity at $y = 0$. However when $0 < s < 1$ and u is smooth and bounded, such integral is well defined as a principal value, that is,

$$\lim_{\varepsilon\downarrow 0}\int_{\mathbb{R}^N\setminus B(0,\varepsilon)}\frac{u(x+y,t)-u(x,t)}{|y|^{N+2s}}dy \tag{C.1.5}$$
$$=\lim_{\varepsilon\downarrow 0}\int_{\mathbb{R}^N\setminus B(x,\varepsilon)}\frac{u(z,t)-u(x,t)}{|z-x|^{N+2s}}dz$$
$$=-\left(C_{N,s}\right)^{-1}(-\Delta)^s u(x,t),$$

for a proper normalizing constant $C_{N,s}>0$ (see (C.1.6) below and Sect. 2.3). This shows that a simple random walk with possibly long jumps produces at the limit a singular integral with a homogeneous kernel. For more details on this topic we refer to [51]. In the case when in (C.1.5), $\mathbb{R}^N$ is replaced by an arbitrary open set $G\subset\mathbb{R}^N$ and the integral kernel is restricted only to the open set, we formally obtain the so-called **regional fractional Laplacian** $-(-\Delta)^s_G$ (cf. [1, 3, 20–22]). More precisely, for

$$u\in\mathcal{L}^1_s(G)=\left\{u:\ G\to\mathbb{R}\ \text{measurable},\ \int_G\frac{|u(x)|}{(1+|x|)^{N+2s}}dx<\infty\right\},$$

and $\varepsilon>0$, we let

$$(-\Delta)^s_{G,\varepsilon}u(x)=C_{N,s}\int_{\{y\in G,|y-x|>\varepsilon\}}\frac{u(x)-u(y)}{|x-y|^{N+2s}}dy,$$

with

$$C_{N,s}=\frac{s2^{2s}\Gamma\left(\frac{N+2s}{2}\right)}{\pi^{\frac{N}{2}}\Gamma(1-s)}, \tag{C.1.6}$$

where Γ denotes the usual Gamma function. Define

$$(-\Delta)^s_G u(x)=C_{N,s}\text{P.V.}\int_G\frac{u(x)-u(y)}{|x-y|^{N+2s}}dy=\lim_{\varepsilon\downarrow 0}(-\Delta)^s_{G,\varepsilon}u(x),\quad x\in G,$$

provided that the limit exists. With the latter definition, the evolution law

$$\partial_t u+(-\Delta)^s_G u=0 \tag{C.1.7}$$

corresponds to a kind of "censored" stable process in $G\subset\mathbb{R}^N$, which is a Lévy motion forced to stay inside G. It is interesting to note that the fractional heat equation (C.1.4) also emerges as the hydrodynamic limit of interacting particle systems that are superdiffusive in nature, that is, the limit of systems on which particles may perform long jumps in the context of Lévy processes [29]. Such "restricted" Lévy motions show up an important models in both applied mathematics and applied probability [1, 3, 9, 25, 29, 35, 46] (cf. also [13–15, 18], on fractional semilinear parabolic equations), as well as in ecological contexts to model the foraging patterns of a variety of organisms (such as, fruit flies, sharks, etc.) [27, 43]. Lévy processes are also well suited to describe turbulent diffusion in rotating flows [48], chaotic phase-transitions in binary systems [10–12], nonlocal transmission phenomena subject to fractional diffusion [16, 17, 32], micelle dynamics and vortex dynamics, and image processing [31].

C.2 Fractional Equation with Nonlocality in Time

Such equations are commonly found in continuum mechanics in the theory of viscoelastic materials [42, Chapter 5], and the theory of heat flows in homogeneous isotropic conductors with fading memory developed between 1960s and 1970s by Coleman, Gurtin, Pipkin and Nunziato (see, for instance, [4, 24, 40]). Let $u = u(x,t)$ denote the (relative) temperature in a rigid body $G \subset \mathbb{R}^N$, at time $t > 0$ and position $x \in G$ and let $e = e(x,t)$ denote the density of internal energy, $q = q(x,t)$ the heat flux, and $r = r(x,t)$ is the external heat supply. According to [4, 24, 40] the balance of energy reads

$$\partial_t e + \operatorname{div}(q) = r, \tag{C.2.1}$$

and we consider the following constitutive relations

$$e = \int_0^\infty e_G(t-\tau)\, u(\tau)\, d\tau, \quad q = -d(u)\nabla u. \tag{C.2.2}$$

Here $d = d(u)$ is a positive smooth function and $e_G \in L^1(\mathbb{R}_+)$ is an internal energy relaxation function that is also assumed sufficiently smooth. Without loss of generality assume $e_G(0) = 0$. An interesting family of models is obtained in the case of heat flows with fading memory of "power-type" when $e_G \equiv g_{1-\alpha}$ for $\alpha \in (0,1)$, g_α is given by (2.1.1). The second equation of (C.2.2) can be recognized as the classical Fourier law for heat flow in G. Let us now further assume that $u(t,x) = u_0(x)$, for all $x \in G$ and $t \leq 0$. We then observe that (C.2.1)–(C.2.2) is equivalent to an equation with a fractional-in-time derivative, of the form

$$\partial_t^\alpha u - \operatorname{div}(d(u)\nabla u) = r. \tag{C.2.3}$$

This easily follows from Definition 2.1.1 in view of the basic computation

$$\begin{aligned} \partial_t e(x,t) &= \partial_t \left(g_{1-\alpha} * u + \int_t^\infty g_{1-\alpha}(\tau)\, u(t-\tau)\, d\tau \right) \\ &= \partial_t \left(g_{1-\alpha} * u + u_0(x) \int_t^\infty g_{1-\alpha}(\tau)\, d\tau \right) \end{aligned}$$

$$= \partial_t (g_{1-\alpha} * u) - u_0 (x) g_{1-\alpha} (t)$$
$$= \partial_t (g_{1-\alpha} * (u - u_0)) (x, t) .$$

Equation (C.2.3) is usually prescribed with a boundary condition for the temperature on the boundary of G, and an initial condition $u (0, x) = u_0 (x)$. Among classical boundary conditions one can take either Dirichlet or Neumann boundary conditions for u. Nonlocal equations of the form (C.2.3) also occur in fluid flows through porous media when the fluid reacts chemically with the medium enlarging the pores and/or obstructing some of the pores. Such problems are of intrinsic interest to geothermic theory. Caputo [2] has proposed to modify the classical empirical law of Darcy by introducing a memory formalism represented by a derivative of fractional order simulating the effect of a decrease of the permeability in time. In this context, the function u has the meaning of fluid pressure and q is related to the fluid mass flow rate per unit volume/area, and $d = d (u)$ stands for the permeability of the porous medium. The classical mass conservation equation of the infiltrating pore fluid is assumed to be

$$\partial_t u + \operatorname{div} (q) = h. \tag{C.2.4}$$

Instead of the classical form of Darcy's law for the mass q, one assumes instead that

$$q := -\partial_t^{1-\alpha}\Big(d (u) \nabla u\Big), \alpha \in (0, 1) , \tag{C.2.5}$$

in order to account for any memory effects present during geothermal flows (see [2]). In view of these considerations, (C.2.4)–(C.2.5) then becomes once again (C.2.3) assuming that $r := \partial_t^{\alpha-1} h$. The fractional kinetic equation (C.2.3) is also dubbed as the equation of *fractional Brownian* motion and arises in various important classes of problems in economics, the study of fluctuations in solids and water flows in hydrology (see [33]; cf. also [37]).

C.3 Space-Time Fractional Nonlocal Equation

Firstly, let us also mention that fractional diffusion equations can be derived from the Continuous-Time Random Walk (CTRW). In fact, a CTRW is a random walk that permits intervals between successive walks to be independent and identically distributed. In this process, the walker starts at the point zero at time $T_0 = 0$ and waits until time T_1 when he makes a jump of size x_1, which is not necessarily positive. He then waits until time T_2 and makes another jump of size x_2, and so on. The jump sizes x_i are assumed to be independent and identically distributed. The intervals $\tau_i = T_i - T_{i-1}, i = 1, 2, \ldots$, are called the *waiting times* and are assumed to be independent and identically distributed. Let T denote the waiting time and X the jump size. Let $f_X(x)$ and $f_T(t)$ denote the PDF of X and the PDF of T, respectively. Let $u(x,t)$ denote the probability that the position of the walker at time t is x, given that it was in position 0 at time $t=0$; that is,

$$u(x,t) = P[X(t) = x | X(0) = 0].$$

Let $R_T(t) := P[T > t]$ be the *survival probability*, which is the probability that the waiting time, when the process is in a given state, is greater that t. If $f_T(t) = \lambda e^{-\lambda t}$ $(t > 0)$, then $R_T(t) = e^{-\lambda t}$ and satisfies the ODE:

$$R_T'(t) = -\lambda R_T(t), \quad t > 0, \quad R_T(0^+) = 1. \tag{C.3.1}$$

A generalization of (C.3.1) that gives rise to anomalous relaxation and power-law tails in the waiting time PDF can be written as

$$\left(g_{1-\alpha} * R_T'\right)(t) = -\lambda R_T(t), \quad t > 0, \quad 0 < \alpha < 1, \quad R_T(0^+) = 1,$$

where we recall that $g_{1-\alpha} * R_T'$ is the classical Caputo fractional derivative of R_T. In that case $R_T(t) = E_{\alpha,1}(-\lambda t^\alpha)$ and the corresponding PDF of the waiting time is $f_T(t) = t^{\alpha-1} E_{\alpha,\alpha}(-\lambda t^\alpha)$. If one assumes that both $f_X(x)$ and $f_T(t)$ exhibit algebraic tails such as $f_X(x) \sim |x|^{-(1+\beta)}$ and $f_T(t) \sim t^{-(1+\alpha)}$, then we can derive a space-time fractional diffusion equation for the dynamics $u(x,t)$ as follows:

$$g_{1-\alpha} * \partial_t u(x,t) = C_{\alpha,\beta} \frac{\partial^\beta u(x,t)}{\partial |x|^\beta},$$

where $C_{\alpha,\beta}$ is a diffusion coefficient. For more details on this topic we refer to the monograph [28, Section 10.7.1].

Secondly, the fractional kinetic equations (C.1.7) and (C.2.3) can be viewed as different limit cases of a more general family of fractional equations, with the first limit being related to Lévy processes (see Appendix C.1) and the second to the problem of fractional Brownian motion assuming that $d \equiv d_0 > 0$ is constant (see Appendix C.2). Both of these two limit cases can be also derived asymptotically from continuous time random walks [36] and can be unified by the following fractional kinetic equation

$$\partial_t^\alpha u + \left(C_{N,s}\right)^{-1} (-\Delta)^s u = 0, \ s \in (0,1), \tag{C.3.2}$$

where as before in Appendix C.1, $u(x,t)$ is the probability that our particle lies at $x \in \mathbb{R}^N$ at time t. Equation (C.3.2) is derived by Montroll-Weiss in [38] using an *integral equation*, the so-called generalized master equation, for processes with time delay that may account for possible trapping of the particles in certain regions before returning to their initial point. It turns out that an equivalence between such master equations and continuous-time random walks (CRWs) can be established (see [30]) by means of an explicit relationship between the (pausing) time distribution function $\psi(t)$ in the theory of CRWs and a memory function ϕ that shows up in the kernel of the master equation. On the basis of the latter equation, there is another probability function $\mathcal{K}(x-\overline{x}) = \mathcal{K}(\overline{x}-x)$ used by the particle to make a step from the position at $x \in \mathbb{R}^N$ to the point $\overline{x} \in \mathbb{R}^N$. Some explicit examples of ψ, ϕ and a transition density $\mathcal{K}$ are presented, for instance, in [45, Section IV], leading to the fractional equation (C.3.2) (cf. also [50], for a more general discussion). Fractional equations of the form

$$\partial_t^\alpha u + (-\Delta)_G^s u = f, \tag{C.3.3}$$

where $f = f(x,t,u)$ is a source, arise as macroscopic transport models for the probability density function of tracer particles in turbulent plasmas [6], which incorporate in a unified way space-time nonlocality. The fractional derivative ∂_t^α accounts for the trapping of tracer particles in turbulent eddies while the diffusion operator $(-\Delta)_G^s$ is responsible for anomalous transport of the tracer particles.

References

1. K. Bogdan, K. Burdzy, Z.-Q. Chen, Censored stable processes. Probab. Theory Rel. Fields **127**(1), 89–152 (2003)
2. M. Caputo, Diffusion of fluids in porous media with memory. Geothermics **28**(1), 113–130 (1999)
3. Z.-Q. Chen, T. Kumagai, Heat kernel estimates for stable-like processes on d-sets. Stoch. Process. Appl. **108**(1), 27–62 (2003)
4. B.D. Coleman, M.E. Gurtin, Equipresence and constitutive equations for rigid heat conductors. Z. Angew. Math. Phys. **18**, 199–208 (1967)
5. P.M. de Carvalho Neto, Fractional Differential Equations: A Novel Study of Local and Global Solutions in Banach Spaces. PhD thesis, Universidade de São Paulo, 2013
6. D. del Castillo-Negrete, B.A. Carreras, V.E. Lynch, Nondiffusive transport in plasma turbulence: a fractional diffusion approach. Phys. Rev. Lett. **94**(6), 065003 (2005)
7. S. Dipierro, X. Ros-Oton, E. Valdinoci, Nonlocal problems with Neumann boundary conditions. Rev. Mat. Iberoam. **33**(2), 377–416 (2017)
8. Q. Du, M. Gunzburger, R.B. Lehoucq, K. Zhou, A nonlocal vector calculus, nonlocal volume-constrained problems, and nonlocal balance laws. Math. Models Methods Appl. Sci. **23**(3), 493–540 (2013)
9. A.A. Dubkov, B. Spagnolo, V.V. Uchaikin, Lévy flight superdiffusion: an introduction. Int. J. Bifur. Chaos Appl. Sci. Eng. **18**(9), 2649–2672 (2008)
10. C.G. Gal, Non-local Cahn-Hilliard equations with fractional dynamic boundary conditions. Eur. J. Appl. Math. **28**(5), 736–788 (2017)
11. C.G. Gal, On the strong-to-strong interaction case for doubly nonlocal Cahn-Hilliard equations. Discrete Contin. Dyn. Syst. **37**(1), 131–167 (2017)
12. C.G. Gal, Doubly nonlocal Cahn-Hilliard equations. Ann. Inst. H. Poincaré Anal. Non Linéaire **35**(2), 357–392 (2018)
13. C.G. Gal, M. Warma, Elliptic and parabolic equations with fractional diffusion and dynamic boundary conditions. Evol. Equ. Control Theory **5**(1), 61–103 (2016)
14. C.G. Gal, M. Warma, Long-term behavior of reaction-diffusion equations with nonlocal boundary conditions on rough domains. Z. Angew. Math. Phys. **67**(4), 42 (2016). Art. 83

15. C.G. Gal, M. Warma, Reaction-diffusion equations with fractional diffusion on non-smooth domains with various boundary conditions. Discrete Contin. Dyn. Syst. **36**(3), 1279–1319 (2016)
16. C.G. Gal, M. Warma, Transmission problems with nonlocal boundary conditions and rough dynamic interfaces. Nonlinearity **29**(1), 161–197 (2016)
17. C.G. Gal, M. Warma, Nonlocal transmission problems with fractional diffusion and boundary conditions on non-smooth interfaces. Commun. Part. Differ. Equa. **42**(4), 579–625 (2017)
18. C.G. Gal, M. Warma, On some degenerate non-local parabolic equation associated with the fractional p-Laplacian. Dyn. Partial Differ. Equ. **14**(1), 47–77 (2017)
19. G. Gripenberg, S.-O. Londen, O. Staffans, *Volterra Integral and Functional Equations*. Encyclopedia of Mathematics and Its Applications, vol. 34 (Cambridge University Press, Cambridge, 1990)
20. Q.-Y. Guan, Integration by parts formula for regional fractional Laplacian. Commun. Math. Phys. **266**(2), 289–329 (2006)
21. Q.-Y. Guan, Z.-M. Ma, Boundary problems for fractional Laplacians. Stoch. Dyn. **5**(3), 385–424 (2005)
22. Q.-Y. Guan, Z.-M. Ma, Reflected symmetric α-stable processes and regional fractional laplacian. Probab. Theory Rel. Fields **134**(4), 649–694 (2006)
23. M. Gunzburger, R.B. Lehoucq, A nonlocal vector calculus with application to nonlocal boundary value problems. Multiscale Model. Simul. **8**(5), 1581–1598 (2010)
24. M.E. Gurtin, A.C. Pipkin, A general theory of heat conduction with finite wave speeds. Arch. Ration. Mech. Anal. **31**(2), 113–126 (1968)
25. R. Gorenflo, F. Mainardi, A. Vivoli, Continuous-time random walk and parametric subordination in fractional diffusion. Chaos Solitons Fract. **34**(1), 87–103 (2007)
26. D. Henry, *Geometric Theory of Semilinear Parabolic Equations*. Lecture Notes in Mathematics, vol. 840 (Springer, Berlin/New York, 1981)
27. N.E. Humphries, N. Queiroz, J.R.M. Dyer, N.G. Pade, M.K. Musyl, K.M. Schaefer, D.W. Fuller, J.M. Brunnschweiler, T.K. Doyle, J.D.R. Houghton, et al. Environmental context explains lévy and brownian movement patterns of marine predators. Nature **465**(7301), 1066 (2010)
28. O.C. Ibe, *Markov Processes for Stochastic Modeling*, 2nd edn. (Elsevier, Inc., Amsterdam, 2013)
29. M. Jara, Nonequilibrium scaling limit for a tagged particle in the simple exclusion process with long jumps. Commun. Pure Appl. Math. **62**(2), 198–214 (2009)
30. V.M. Kenkre, E.W. Montroll, M.F. Shlesinger, Generalized master equations for continuous-time random walks. J. Stat. Phys. **9**(1), 45–50 (1973)
31. J. Klafter, M.F. Shlesinger, G. Zumofen, Beyond brownian motion. Phys. Today **49**(2), 33–39 (1996)
32. D. Kriventsov, Regularity for a local-nonlocal transmission problem. Arch. Ration. Mech. Anal. **217**(3), 1103–1195 (2015)
33. B.B. Mandelbrot, J.W. Van Ness, Fractional Brownian motions, fractional noises and applications. SIAM Rev. **10**, 422–437 (1968)
34. A.M. Mathai, H.J. Haubold, Special functions for applied scientists. 4, 2008. https://doi.org/10.1007/978-0-387-75894-7
35. A. Mellet, S. Mischler, C. Mouhot, Fractional diffusion limit for collisional kinetic equations. Arch. Ration. Mech. Anal. **199**(2), 493–525 (2011)
36. R. Metzler, J. Klafter, The random walk's guide to anomalous diffusion: a fractional dynamics approach. Phys. Rep. **339**(1), 77 (2000)
37. E.W. Montroll, M.F. Shlesinger, On the wonderful world of random walks, in *Nonequilibrium Phenomena, II*. Studies in Statistical Mechanics, vol. XI (North-Holland, Amsterdam, 1984), pp. 1–121
38. E.W. Montroll, G.H. Weiss, Random walks on lattices. II. J. Math. Phys. **6**, 167–181 (1965)
39. R. Muralidhar, D. Ramkrishna, H. Nakanishi, D. Jacobs, Anomalous diffusion: a dynamic perspective. Phys. A: Stat. Mech. Appl. **167**(2), 539–559 (1990)

40. J.W. Nunziato, On heat conduction in materials with memory. Quart. Appl. Math. **29**, 187–204 (1971)
41. I. Podlubny, *Fractional Differential Equations*. Mathematics in Science and Engineering, vol. 198 (Academic, San Diego, CA, 1999). An introduction to fractional derivatives, fractional differential equations, to methods of their solution and some of their applications
42. J. Prüss, *Evolutionary Integral Equations and Applications*. Modern Birkhäuser Classics (Birkhäuser/Springer Basel AG, Basel, 1993) [2012]. Reprint of the 1993 edition
43. A.M. Reynolds, C.J. Rhodes, The lévy flight paradigm: random search patterns and mechanisms. Ecology **90**(4), 877–887 (2009)
44. F. Rothe, *Global Solutions of Reaction-Diffusion Systems*. Lecture Notes in Mathematics, vol. 1072 (Springer, Berlin, 1984)
45. A.I. Saichev, G.M. Zaslavsky, Fractional kinetic equations: solutions and applications. Chaos **7**(4), 753–764 (1997)
46. D. Schertzer, M. Larchevêque, J. Duan, V.V. Yanovsky, S. Lovejo, Fractional Fokker-Planck equation for nonlinear stochastic differential equations driven by non-Gaussian Lévy stable noises. J. Math. Phys. **42**(1), 200–212 (2001)
47. W.R. Schneider, Grey noise, in *Stochastic Processes, Physics and Geometry (Ascona and Locarno, 1988)* (World Scientific Publishing, Teaneck, NJ, 1990), pp. 676–681
48. T.H. Solomon, E.R. Weeks, H.L. Swinney, Observation of anomalous diffusion and lévy flights in a two-dimensional rotating flow. Phys. Rev. Lett. **71**(24), 3975 (1993)
49. S. Umarov, R. Gorenflo, On multi-dimensional random walk models approximating symmetric space-fractional diffusion processes. Fract. Calc. Appl. Anal. **8**(1), 73–88 (2005)
50. M. Vahabi, B. Shokri, Transport in inhomogeneous systems. Rep. Math. Phys. **69**(2), 243–250 (2012)
51. E. Valdinoci, From the long jump random walk to the fractional Laplacian. Bol. Soc. Esp. Mat. Apl. SeMA 49, 33–44 (2009)
52. L. Vlahos, H. Isliker, Y. Kominis, K. Hizanidis, Normal and anomalous diffusion: a tutorial (2008). Preprint. arXiv:0805.0419
53. M. Warma, A fractional Dirichlet-to-Neumann operator on bounded Lipschitz domains. Commun. Pure Appl. Anal. **14**(5), 2043–2067 (2015)
54. M. Warma, The fractional relative capacity and the fractional Laplacian with Neumann and Robin boundary conditions on open sets. Potential Anal. **42**(2), 499–547 (2015)
55. M. Warma, The fractional Neumann and Robin type boundary conditions for the regional fractional p-Laplacian. Nonlinear Differ. Equ. Appl. **23**(1), 46 (2016). Art. 1

Index

C. G. Gal, M. Warma, *Fractional-in-Time Semilinear Parabolic Equations and Applications*, Mathématiques et Applications 84,
https://doi.org/10.1007/978-3-030-45043-4

P

Q

R

S

U

W

Printed in the United States
By Bookmasters